Incentive-Centric Semantic Web Application Engineering

Synthesis Lectures on the Semantic Web: Theory and Technology

Editors
James Hendler, *Rensselaer Polytechnic Institute*
Ying Ding, *Indiana University*

Synthesis Lectures on the Semantic Web: Theory and Application is edited by James Hendler of Rensselaer Polytechnic Institute. Whether you call it the Semantic Web, Linked Data, or Web 3.0, a new generation of Web technologies is offering major advances in the evolution of the World Wide Web. As the first generation of this technology transitions out of the laboratory, new research is exploring how the growing Web of Data will change our world. While topics such as ontology-building and logics remain vital, new areas such as the use of semantics in Web search, the linking and use of open data on the Web, and future applications that will be supported by these technologies are becoming important research areas in their own right. Whether they be scientists, engineers or practitioners, Web users increasingly need to understand not just the new technologies of the Semantic Web, but to understand the principles by which those technologies work, and the best practices for assembling systems that integrate the different languages, resources, and functionalities that will be important in keeping the Web the rapidly expanding, and constantly changing, information space that has changed our lives.
Topics to be included:

- Semantic Web Principles from linked-data to ontology design

- Key Semantic Web technologies and algorithms

- Semantic Search and language technologies

- The Emerging "Web of Data" and its use in industry, government and university applications

- Trust, Social networking and collaboration technologies for the Semantic Web

- The economics of Semantic Web application adoption and use

- Publishing and Science on the Semantic Web

- Semantic Web in health care and life sciences

Incentive-Centric Semantic Web Application Engineering
Elena Simperl, Roberta Cuel, and Martin Stein

ISBN: 978-3-031-79440-7   paperback
ISBN: 978-3-031-79441-4   ebook

DOI 10.1007/978-3-031-79441-4

A Publication in the Springer series
SYNTHESIS LECTURES ON THE SEMANTIC WEB: THEORY AND TECHNOLOGY

Lecture #4
Series Editors: James Hendler, *Rensselaer Polytechnic Institute*
                Ying Ding, *Indiana University*
Series ISSN
Synthesis Lectures on the Semantic Web: Theory and Technology
ISSN pending.

Incentive-Centric Semantic Web Application Engineering

Elena Simperl
University of Southampton, United Kingdom

Roberta Cuel
University of Trento, Italy

Martin Stein
University of Siegen, Germany

SYNTHESIS LECTURES ON THE SEMANTIC WEB: THEORY AND TECHNOLOGY #4

ABSTRACT

While many Web 2.0-inspired approaches to semantic content authoring do acknowledge motivation and incentives as the main drivers of user involvement, the amount of useful human contributions actually available will always remain a scarce resource. Complementarily, there are aspects of semantic content authoring in which automatic techniques have proven to perform reliably, and the added value of human (and collective) intelligence is often a question of cost and timing. The challenge that this book attempts to tackle is how these two approaches (machine- and human-driven computation) could be combined in order to improve the cost/performance ratio of creating, managing, and meaningfully using semantic content.

To do so, we need to first understand how theories and practices from social sciences and economics about user behavior and incentives could be applied to semantic content authoring. We will introduce a methodology to help software designers to embed incentives-minded functionalities into semantic applications, as well as best practices and guidelines. We will present several examples of such applications, addressing tasks such as ontology management, media annotation, and information extraction, which have been built with these considerations in mind. These examples illustrate key design issues of incentivized Semantic Web applications that might have a significant effect on the success and sustainable development of the applications: the suitability of the task and knowledge domain to the intended audience, and the mechanisms set up to ensure high-quality contributions, and extensive user involvement.

KEYWORDS

semantic content creation, ontology engineering, media annotation, information extraction, Web service annotation motivation, incentives, mechanism design, participatory design, games with a purpose, gamification

Contents

Preface

In the last few years, the Semantic Web community, academia, as well as industry, developed a great amount of technology to create, manage, and use semantic content. These tools are, however, hardly accessible to a lay audience, mainly because they require in-depth expertise not only with respect to the task at hand, but most importantly with respect to the underlying processes and procedures according to which the tasks are executed. These limitations laid the foundations for the emergence of a new field of research in semantic technologies that attempted to draw from the success of Web 2.0 to encourage user participation and make use of the power of collective intelligence. The basic idea is to study how specific tasks in the semantic data management life cycle could be optimally approached as a combination of human and computational intelligence. In particular, the underlying applications should by design contain features and implement interaction concepts that motivate users to participate at the resolution of the task—for instance, to collaboratively create useful ontologies, identify missing, inconsistent or repeating data entries, and classify content according to a given classification scheme.

In order to analyze the conditions under which a user—either in isolation or working with others—will actively contribute toward the completion of a task, we need to look at scientific results in social sciences and economics on motivations and incentives. Especially in organizational management, there are various theories and case studies on how to trigger and influence human behavior to achieve specific aims, and these insights could and should be applied to the field of semantic technologies. Closer to the actual development of IT systems, there is an impressive body of knowledge in Human Computer Interaction (HCI), most concretely in the field of participatory design, which is concerned with how users can be engaged in the very design of the applications they are expected to interact with, thus increasing the acceptance and level of usage of these applications. Such participatory-design methods could be of great benefit to Semantic Web developers who target scenarios with large user involvement.

In this book we discuss how this could be achieved. In Chapter 1 we analyze core semantic data management activities in light of their potential to be feasibly automated, and identify those tasks for which human input is indispensable. The applications used as examples should give a first idea of how such human inputs could be encouraged, and of the type of questions an incentives-minded application design should consider in order to achieve the application goal through contributions of its user base. In Chapter 2 we introduce a range of methods from mechanism design, organizational management, and user experience design that can be applied in real-world projects in order to not only trigger participation, but also steer user behavior toward the completion of particular tasks at a feasible level of quality and costs. Chapters 3–5 describe case studies using these methods in the areas of enterprise knowledge management, Web service annotation and games with a purpose for knowledge creation. Chapter 6 provides a summary of best practices and lessons learned in the case studies reported in the previous chapters.

This book is a result of the European research project Insemtives, funded by the 7th Framework Programme under the grant number FP7-231181. The authors would like to thank the Insemtives team for the fruitful collaboration throughout the three years of the project, which played an important role in the realization of this book.

Elena Simperl, Roberta Cuel, and Martin Stein
January 2013

CHAPTER 1

Semantic Data Management: A Human-driven Process

1.1 FUNDAMENTALS OF SEMANTIC DATA MANAGEMENT

The Semantic Web refers to a set of standards and technologies that are used to enable the machine-driven design and operation of a wide range of aspects of data management, from the description and organization of collections of digital artifacts, to the access to distributed digital repositories, as well as search, ranking, and recommendation. In all these areas, which may be relevant to different vertical domains and usage scenarios, semantics provides the tools to achieve a higher degree of automation of the underlying computational tasks, thus enabling scalability. In addition, the usage of semantic standards and technologies alters the ways the corresponding applications are developed, and how information is created and evolves as part of these applications; in particular, semantics facilitates information reusability, and mediation across distributed information spaces, which are discovered and explored at run time—independently of the conceptualization they use to describe the underlying information, and their means of (internal) organization.

The core features of semantically enabled applications could be summarized as follows:

- Data is self-described. It is encoded as RDF (Resource Description Framework) using a triple-based data model and accessed via Web standards and protocols such as URIs (Uniform Resource Identifier) and HTTP (Hypertext Transfer Protocol).

- The meaning of data is specified using ontologies. Ontologies capture the types of things of interest in a given vertical domain, as well as their main attributes and relationships. They are used to augment the way machines can manage and use information artifacts, for instance by processing semantic markup of HTML pages or dedicated repositories containing annotations of multimedia content. Semantic Web ontologies are encoded in languages such as RDFS (RDF Schema) and OWL (Web Ontology Language).

- Reasoning can be used to infer new facts from a Semantic Web knowledge base, to classify new information according to an ontological schema, and to identify potential inconsistencies and other mismatches in data quality.

- In a distributed scenario different ontological conceptualizations may co-exist. Providing a unified access to decentralized knowledge bases is facilitated by the availability of links between individual ontologies and instance data sets. These links capture similarities between entities described according to the different conceptualizations. Processing and managing data in a distributed, global space cannot be predicated by the availability of the underlying schemas at design time; data management services are expected to operate in an open environment, where new data sources can be discovered, integrated, and used on the fly.

• Closed-world scenarios can equally take advantage of semantic technologies. Typically, in such cases one builds (or reuses) domain-specific ontologies represented using Semantic Web standards, applies reasoning techniques to automatically deduce new facts and identify conceptual or classification errors, and uses the resulting knowledge base to provide intelligent—as in, aware of the meaning of the domain and user intentions in the given application context—information management capabilities, including information organization, retrieval, and recommendation.

Automation—as a means to deal with scale and optimize development and maintenance costs—is achieved at various levels along the data management life cycle; it involves the creation of new semantic data (instances as well as ontological schemas, possibly from existing legacy sources), its evolution (responding to different feedback channels and changes in the knowledge domain), and use in various application scenarios.

Each and every one of these aspects has been subject to intensive research over the past ten years or more. This led to numerous automatic and semi-automatic methods, techniques, and tools that support the data management life cycle, complemented by their manual counterparts, which provide process assistance and interfaces abstracting from language technicalities to knowledge engineers, application developers, and data owners. For example, we now have ontology development environments such as Protégé that facilitate the (collaborative) creation and management of ontologies; methods to learn and populate ontologies by extracting knowledge from unstructured and semi-structured resources such as text documents; methods to automatically discover similar entities within different RDF data sets; content management systems supporting the creation of semantic markup; reasoners operating over large-scale Semantic Web knowledge bases containing billions of RDF triples; and query engines reaching over the entire Web of Data to deliver answers to user-specific information needs, organizing data spaces according to the meaning of the underlying content, and offering novel means to access and explore them.

Due to the knowledge-intensive nature of these activities, most tools, however, cannot be operated in fully automated fashion. Human intervention remains a core part of the data management life cycle, even if it merely involves the configuration of machine-driven computational tasks through the creation of training examples or the validation of automatically computed results. In the remainder of this chapter we will shed light on the hybrid nature of semantic data management; we will visit the main phases of the data management life cycle, identify those aspects which fundamentally rely on human inputs, and characterize and classify the various types of inputs that are needed. The subsequent chapters, in particular Chapters 3 to 5, will illustrate how semantically enabled applications could be designed in order to engage with their users more closely, and acquire the Semantic Web data that is required to cover these aspects.

1.2   CREATING, MANAGING, AND USING SEMANTIC DATA

1.2.1   OVERVIEW OF THE SCENARIOS

The semantic data management life cycle can be briefly described as a series of activities involving the creation, maintenance, and usage of semantic data, ontologies, and instances populating them. Sometimes, these activities are undertaken in parallel or overlap; for instance, using a semantic application might give hints to changes required at the data level, which in turn leads to maintenance. Activities may take different forms; most notably, the creation of semantic data can be carried out in a purely manual fashion, informed by a variety of domain-specific resources, or directly leverage such resources to translate them into RDF and ontological structures.

In this section, we will introduce several semantic data management scenarios, which can be found in a great majority of real-world Semantic Web projects; they will form the basis for a subsequent analysis of the extent to which they require human contributions, and of the crowdsourcing approaches they are likely to be amenable to. Our scenario descriptions follow an ontology-centric view on the overall data management life cycle; alternatively, one could put a stronger emphasis on the actual data—as it is prominently promoted by the Linked Data initiative [54]—and see ontologies as means to structure and further describe it. Both perspectives are equally valid, and yield similar results in terms of the goals of our endeavor—that is, to identify where human intelligence should be employed in order to optimize semantic data management methods, techniques, and tools.

One of the core components of semantic data management life cycle are *ontologies*: shared specifications of conceptualizations of a domain [49]. Ontology engineering is a challenging process that encompasses a broad range of tasks and activities, as illustrated in Figure 1.1 [44].

Ontology management refers primarily to scheduling, controlling, and quality assurance. Scheduling is about coordinating and managing an ontology development project, including resource and time management. Controlling ensures that the scheduled tasks are accomplished as planned. Finally, quality assurance evaluates the quality of the outcomes of each activity, most notably of the implemented ontology.

Ontology development can be split into three phases: pre-development, development, and post-development. As part of the pre-development phase, an environment study investigates the intended purpose and use of the ontology. Furthermore, a feasibility study ensures that the ontology can actually be built within the time and resources assigned to the project. These two activities are followed by the actual development, which includes first and foremost the requirements specification that eventually results in a conceptual model and its implementation in a given knowledge representation language. In the final, post-development phase, the ontology is updated and maintained as required; this phase also includes the reuse of the ontology in other application scenarios.

Support stands for a wide range of different activities that can be performed in parallel or subsequent to the actual ontology development. The aim of these activities is to augment the results of the (typically manual) ontology development by automatizing parts of the process, providing auxiliary information sources that could be used to inform the conceptualization and implementation tasks,

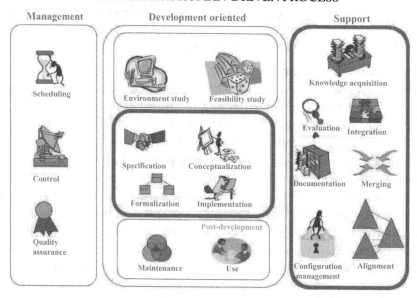

Figure 1.1: Ontology engineering activities [44].

and evaluating and documenting intermediary results. Typical support activities include *knowledge acquisition, ontology evaluation, ontology alignment*, and *ontology learning* and *ontology population*.

Ontology population is closely related to *semantic annotation*, by which information artifacts of various forms and flavors are described through instances of a given ontology. Instance data can be created by *RDFizing* structured or semi-structured sources such as CSV files, data bases, glossaries, catalogues, or by extracting entities from text. As part of this process, the data provider may decide to reuse an existing ontology, and translate the legacy structures to the target, Semantic-Web-centric representations. Alternatively she may use the (implicit or explicit) structures according to which the original data sources were stored and managed as a baseline for a new Semantic Web ontology, to which the newly created RDF data refers to. The Linked Data initiative provides best practices and recommendations for the design and technical realization of this process, including the definition of identifiers for RDF resources, the reuse of existing ontologies, the online publication of the resulting RDF data set, and the interlinking to other Linked Data data sets.

Data interlinking is closely related to the area of *ontology alignment*, and involves the definition of correspondences between entities located in different RDF data sets published according to Linked Data principles, and the description of these correspondences through specific predicates (equivalence, related to, or domain-specific ones). The two activities not only share commonalities in terms of the types of basic (machine-driven) algorithms they make use of, but can also influence each other. Based on mappings at the schema level, one can identify potentially related instances; the other way around, the availability of links between sets of entities may indicate similarities between classes.

In the following we look into more detail at these scenarios to identify lower-level tasks and analyze the extent to which they can be feasibly automated. In this context, feasibility refers primarily to the trade-off between the effort associated with the usage of a given tool targeting automation—including aspects such as getting familiar with the tool, but more importantly creating training data sets and examples, configuring the tool and validating (intermediary) results—and the quality of the outcomes. The fundamental question is to identify patterns of hybrid data management workflows, in which human and machine-driven computation are combined in a way that yields optimal results, both in terms of costs in the short, medium, and long term, but also with respect to accuracy. As such, it is clear that on the effort side, aspects related to system usability are likely to play a role in this trade-off analysis. The ability of users—experts or laymen—to handle a given data management task via a tool is influenced by the extent to which the tool is built in a user-friendly manner. This book does not focus on user-centric aspects of the Semantic Web, as treated, for example, in [84, 91, 101, 108, 130], but on specific principles and best practices to attract user contributions. In this context, Chapter 4 considers user experience design from a participatory perspective, as a means to achieve a stronger user engagement by involving users in the design process of semantic tools and applications.

We now turn to the analysis of the semantic data management activities discussed earlier. The analysis is a summary of a previous work of ours [115], which gives full particulars on the methodologies, methods, and tools covering each activity. We surveyed them in order to learn about the types of processes semantic data management projects conform to, and the extent to which and reasons they might rely on human intervention.

1.2.2 DEVELOPING ONTOLOGIES

Developing ontologies requires domain expertise and the ability to capture domain knowledge in a clean conceptual model. An ontology describes the things that are important in a specific domain of interest, their properties, and the way they are interrelated. It defines a common vocabulary and the meaning of the terms used in the vocabulary. In the last decade, several ontology development methodologies have been proposed [44]. Many suggest to start with the specification of the scope the ontology should cover and the requirements it should fulfill. This is often complemented by the informal and formal specification of competency questions. Based on that, relevant terms in the domain are then collected. Widely accepted ontology representation formalisms such as RDFS and OWL use classes, properties, instances, and axioms as ontological primitives to describe domain knowledge. Ontologies can vary in their degree of formality, ranging from highly informal (defined in natural language), to semi-informal (the language is restricted or controlled), semi-formal (formally defined language), to rigorously formal [44]. The overall process can be performed in a centralized (within a pre-defined team of knowledge engineers and domain experts) or a decentralized fashion (within a potentially open community of stakeholders, domain experts, and users).

Describing the Domain and Scope of the Ontology

This task results in a thorough description of what the ontology is expected to cover. It outlines the domain, coverage, and focus of the ontology and lists the intended users, and usage scenarios. The task can also involve defining requirements that the ontology must or should fulfill. In collaborative ontology engineering, an important aspect of the task is to establish consensus among the different stakeholders in terms of the requirements to be fulfilled. As in many other engineering disciplines, requirements analysis and gathering is mostly a manual process.

Specifying Competency Questions

This task can be seen as a continuation of the previous one: competency questions are a means to further specify the domain and scope of the ontology, and to evaluate the ontology in a later stage. They capture the types of questions or queries that the ontology is expected to answer based on the identified usage scenarios. Competency questions can be specified in a formal manner. Collecting competency questions that reflect the ontology engineering requirements is a challenging conceptual task. This also applies to the formalization of such questions in a language with a machine-understandable semantics. In both cases human input is crucially required.

Reusing Existing Ontologies

Based on the state of the art, automation support for the reuse of existing ontologies seems feasible when it comes to finding potential reuse candidates. The evaluation and selection of such candidates is a more challenging task, hardly specifiable in great detail from a procedural perspective, and thus difficult to automate [16]. The situation is slightly different for the customization of the relevant ontologies to the characteristics of the new application setting. For this type of task one could resort to tools for extracting ontology fragments, translating to different knowledge representation formalisms, and ontology alignment, though not all are operable in a fully automatic fashion. The same holds for the reuse of related knowledge structures such as taxonomies, classification systems, or relational schemas, where an initial ontology can be extracted from these structures by following generic mapping patterns taking into account the types of modeling primitives they support and their meaning, but the results are typically subject to manual post-processing.

Defining the Vocabulary

For the collection of relevant terms in an ontology Uschold and King propose three different strategies: bottom-up, top-down, and middle-out [131]. Bottom-up means to start at a finer-grained level of specificity and to generalize to broader terms. Top-down is the opposite approach, in which abstract concepts are identified first and are then specialized into more detailed, narrower concepts. Finally, middle-out is a combination of the previous two, in which the most relevant terms are selected first, regardless their level of abstraction, in order to specialize or generalize as required. The outcome of this task is the vocabulary used by the ontology.

Automatic support for vocabulary definition has proven feasible at a basic level. There are tools that process a set of relevant documents and extract terms that occur often [99]. However, the question of how to actually model these terms in the ontology—for example, as class or instance, or as class or property—is highly contextual, and depends on human input. Automation would be possible only to the extent that certain assumptions are expected to hold—for instance, all nouns are supposed to be modeled as classes; subsequently, these assumptions need to be manually validated on a case-by-case basis.

Conceptual Modeling and Knowledge Representation
The goal of this task is to decide on the typing of the terms identified in the previous task. This includes the definition of classes and the associated hierarchy, as well as the definition of properties and additional axioms. Several automatic approaches have been proposed to discover specific types of relationships, in particular specialization and generalization, from natural language text, but human intervention is required for training the underlying algorithms, building the text corpus on which they operate, and validating their results. In addition, efforts need to be typically invested in post-processing the domain and ranges of individual properties, so that these are defined at the most appropriate level in the abstraction hierarchy.

Defining axioms, on the other side, involves specifying precise, logics-based rules, such as cardinality constraints and disjointness that apply to classes of entities. Approaches for automatically specifying such axioms are very limited in their scope and require substantial training and validation [135].

The creation of instances is related to semantic annotation; we investigate it in detail in the next section. Relevant for the context of ontology development is the definition of so-called "fixed" or "ontological" instances which are the result of explicit modeling choices during the conceptualization phase. The distinction between classes and instances is very specific to the application setting, and we are not aware of any approaches aiming at automatizing this task.

1.2.3   CREATING INSTANCE DATA
Creating semantic data as instances of an ontology can take at least two different forms: annotating digital artifacts using machine-understandable descriptions represented as RDF and referring to ontological concepts, and publishing data (on the Web) according to Linked Data principles.

Semantic Annotation
There is a wide range of approaches that carry out semi-automatic annotation of texts: most of them make use of natural language processing and information extraction techniques. Even though they require training, a large share of the work can be automated [105, 129]. The situation is slightly different with the annotation of multimedia content: approaches for the annotation of media, whether

manual, semi-automatic, or automatic, aim at closing the so-called "semantic gap," which is a term coined to describe the discrepancy between low-level technical features of multimedia, which can be automatically extracted to a great extent, and the high-level, meaning-bearing features a user is typically interested in and refers to when searching for content. Recent research in the area of semantic multimedia retrieval attempts to derive meaning from low-level features, or other available basic metadata, automatically. This can so far be achieved to a very limited extent, i.e., by applying machine learning techniques with a vertical focus for a specific domain (such as face recognition), in turn for a substantial training and tuning, all undertaken with human intervention [11]. The annotation of Web services is currently a manual task, but more research is needed in order to clearly determine whether this can be traced back to the nature of the task, or to the fact that the corresponding area is not mature enough to produce approaches that can offer reliable automatic results [30, 69, 70].

In [115] we analyzed various tools for text and media annotation which create semantic metadata with respect to the degree of automation they can support (nine tools in the first category, and six in the second one). In the case of textual resources, the main challenge is in finding optimal ways to integrate human inputs (both workflow-wise and implementation-wise) with existing pre-computed results. On the contrary, multimedia annotation remains largely unsolved; there the typical scenario would use human computation as a main source of input for the creation of annotations, though specific optimizations of the process are nevertheless required. Chapter 5 gives examples of mechanisms which lead to higher accuracy and better labor management within games with a purpose for media annotation.

Creating Linked Data

The main tasks that have to be performed in order to publish Linked Data on the Web are (i) to transform the original sources to RDF, possibly resorting to existing ontologies and vocabularies as ground structure of the data; (ii) to assign consistent URIs to the data to be published; (iii) to generate links to other resources exposed as Linked Data; and (iv) to publish metadata which allows further exploration and discovery of the data. Various aspects of these tasks can be performed using state-of-the-art technology in an automated fashion. For instance, tools such as D2R Server and Virtuoso produce RDF from relational databases, but require manual input for the definition of mapping rules to guide the overall process [54]. Automated interlinking is approached through tools such as Silk[136], LIMES [94], and EAGLE [95]. Human intervention is needed to define general correspondence patterns, and to train and adjust the outputs of the underlying algorithms. Tools such as Sindice[1] facilitate the discovery of candidate resources to which the newly published data set can link to, however the actual choice of the target data sets and individual resources is human-guided due to its heavily contextual nature.

---

[1]http://sindice.com/

1.2.4   SUPPORTING ONTOLOGY DEVELOPMENT

Support activities accompany the development of ontologies. One prominent example thereof is *ontology learning*, which can be understood as the (semi-)automatic support to a foremost manual ontology building [85]. This means that knowledge is acquired and classified as specific ontological primitives (typically classes, instances or named entities, relations) from various heterogeneously structured sources. Most approaches to ontology learning require human input at least at run time, and expert insight for configuration and training. In [115] we analyzed several process-oriented methodologies published in the literature of the last years: the execution of ontology learning tools can be done automatically, even though the tools require human input and feedback in key phases of the process. The remaining tasks, such as preparation and selection of appropriate tools, are inconceivable without a significant amount of domain and technology expertise. The variety of tools available makes a general assessment difficult. However, as presented in [115], seven out of 17 tools need user intervention throughout the entire process, eight can be run semi-automatically, and only two are fully automated. Putting aside the actual handling of the tools, collective intelligence may prove of use in defining domain-specific rules for the extraction of ontological knowledge, and for training and validation.

Another important support activity is the *alignment* of heterogeneous ontologies. Many of the existing ontology engineering environments provide means for the manual definition of mappings between ontologies. In addition, there is a wide range of algorithms that provide automatic support [34, 35, 93, 96], while it is generally accepted that the question of which ontological primitives match cannot (yet) be done fully automatically [35, 36]. This area is closely related to *data interlinking*, which we analyzed in more detail in [112, 144].

Ontology evaluation is a very broad topic. In its nature it is still human-driven as it has to either evaluate what was initially produced manually in the conceptual modeling phase, or the usability of an existing ontology against a new set of requirements. Automatic support is available for checking basic structure-based features of taxonomies [43], while front-end tools are helpful to guide the human evaluator throughout the sometimes complex evaluation procedure. Similarly, to the requirements analysis or the conceptualization, ontology evaluation is hardly amenable to automation for generic domains and application settings, as it requires in depth knowledge in ontology engineering [18, 50, 83, 100] that cannot be recorded in a generic tool.

The *documentation* of an ontology is an essential component of ontology engineering to facilitate maintenance and reuse, and to ensure a smooth operation of a collaborative process. Documentation remains human-driven, especially when it comes to recording modeling decisions and their rationales. Basic support for ontology documentation can be obtained by automatically creating entries for each ontological primitive which capture its core context in terms of labels and other annotations, as well as related classes, instances and properties. In this context, it is also worth mentioning the topic of *ontology localization*, which mainly refers to the translation of labels to different natural languages. Similarly, to other areas in ontology engineering which employ natural language processing techniques—for instance, ontology learning—human input is required in order to solve translation questions which are highly context-specific, or to choose between different alternative translations.

In summary, automatic approaches to semantic data management activities have major limitations. As the current state of the art demonstrates in cases which are knowledge-intensive or highly contextual advances are conceivable only at the expense of low accuracy and manual post-processing overhead. Even more, this applies foremost for data sources which are of textual nature—including numerical data. For multimedia, automated support is feasible only for very specific tasks, such as the recognition of a given type of object, and in general the tasks are much more challenging to be solved by a machine than by people. Independently of the actual data, one can distinguish between different forms of human contributions: those required to operate a given tool, which will not be considered further in this book, and the actual domain knowledge that can be acquired from various communities of users, including a designated team of domain experts and the so-called open "crowd." The latter is of particular interest for the remaining chapters; we will look into means to create incentives that stimulate users of a semantic application to participate in semantic data management tasks that rely on domain knowledge. As outlined earlier in this chapter, these tasks typically involve solving questions such as classifying or describing a digital artifact, such as an image, according to a given set of classes defined by an ontology; determining whether a certain type of relationship holds between two classes; confirming that two entities are similar or the same; and so on. From the set of individual answers users provide to questions like these—possibly responding to incentive mechanisms embedded in the application—the application designer derives the actual input that is needed to complete the task, possibly by further feeding this input into a tool pipeline. The next section gives an overview of the main classes of motivators and incentives which have been empirically proven to have a positive effect on the amount and quality of human contributions to the resolution of such tasks. Chapter 2 introduces methods and techniques by which the application designer can identify and study the effects of different models and combinations of such incentives in the context of the application.

1.3   ATTRACTING HUMAN CONTRIBUTIONS

In the analysis of the conditions under which a person—either in isolation or working with others—will actively contribute toward the completion of a task, the literature in social sciences and economics distinguishes between external and internal motivations—that is, socially derived and individually based motivations (see Chapter 2). This distinction is primarily related to the relationship between the task to be undertaken and the willingness of the subject to undertake it; in this respect, one speaks about "intrinsic" and "extrinsic" motivations, where the first refers to tasks that tend to be rewarding in itself, and the second to those where external rewards are the main driver behind their completion.

In the last few years researchers from different scientific disciplines investigated the grounds of the success of community-driven content creation. Although these studies reveal that the inner motivations that drive people to participate are heterogeneous and strongly context-specific, several main categories can be identified:

Reciprocity and expectancy Reciprocity means that contributors receive an immediate or long-term benefit in return for spending their time and resources on performing a certain task. An example of this is tagging, where the user organizes her knowledge assets, such as bookmarks or pictures, and while doing so, reuses tag classifications of other users.

Reputation Reputation is an important factor within (virtual) communities: it can drive users to invest time and effort to improve the opinion of their peers about them. It has shown that this is an important motivation for, for instance, Wikipedia users [75].

Competition Competition is a relevant incentive in the context of games (rankings), but it can also be a strong driver in community portals where the user with the, for example, most contributions, or highest status is awarded.

Conformity to a group Through imitation and empathy people tend to conform to the social group they belong to, therefore making available information about members behaviors is a way to spur people to act according to this information. Staying with the example of Wikipedia as a strong community, studies have shown that the feeling of belonging to a group makes users be more active and supportive. They feel needed and thus obliged to contribute to the joint goals of their community [75].

Altruism People contribute to a joint endeavor because they believe it is for a good cause, without expecting anything in return.

Self-esteem and learning People contribute to a certain endeavor in order for them to grow as individuals, either in terms of their own self-perception or to increase their knowledge and skills.

Fun and personal enjoyment People get engaged in entertaining activities. The most popular approach in this context hides a complex task behind casual games [114, 132].

Implicit promise of future monetary rewards People typically act in order to increase their own human capital and to position themselves to gain future rewards (money or better roles).

Rewards People receive a direct compensation which can obviously play a large role in explaining the rationale for contributing effort toward a goal. Examples of this are crowdsourcing platforms, such as Amazon's Mechanical Turk[2] or InnoCentive.[3]

To identify the motivations that better spur humans to contribute, the first thing an application developer should do is to understand the inner interests that drive users to contribute and the social environment in which contributors act [24]. These interests are affected by various elements:

[2]https://www.mturk.com
[3]http://www.innocentive.com

- The goal of the application and the reasons for user contributions and participation.

- The nature of the good produced through user contributions and the way it is used or consumed (public, private, or club good).

- The actual tasks that users are expected to perform.

- The skills, competencies, and abilities required to undertake the tasks.

- The (simplified) set of social relationships among the subjects participating in the exercise.

More details about the rationales for the choice of these elements, and the methods and techniques from social and economical sciences application designers can use to gain more insight into their instantiation in a given application context are available in Chapter 2.

1.4   EXAMPLES OF INCENTIVIZED SEMANTIC WEB APPLICATIONS

In this section we will give some examples of end-user applications that leverage data described using Semantic Web standards and apply specific types of incentives in order to acquire and maintain these semantic descriptions.

1.4.1   THE SOCIAL SEMANTIC WEB

The field referred to as the "Social Semantic Web" was triggered by the advent of Web 2.0. Creating a social novelty, rather than a technical one, Web 2.0 marked a new era of usability and user participation, virally becoming a worldwide phenomenon that allowed people from all over the world to connect and collaborate on various topics.

New tools and technologies emerged that allowed an increased degree of interaction and exchange not only between users, designers, and providers, but also within the user community. Wiki-based systems, such as Wikipedia, and social tagging platforms such as del.icio.us[4] or Flickr[5] are only a few of the most prominent representatives of this innovative social and technical development. They provide incentives to their users to become engaged with the system, leading to rapidly growing, useful amounts of content [55, 75, 88]. The quality of the content is a consequence of the process through which it is created. As it is built from contributions from a wide range of individuals and organizations, it leverages the wisdom of the crowds, being constantly refined, improved, and updated. At the same time, it reflects the views of its users and the ways the actual system is being used, leading to a greater engagement and straightening the links between application system and its user base.

The overlap between Web 2.0 and the Semantic Web led to exciting developments. On the one side, content produced on Web 2.0 platforms was used to create semantic data, which bears

---

[4]http://del.icio.us
[5]http://www.flickr.com

a more refined structure and formal meaning; this can be best exemplified through the conversion of "folksonomies," vocabularies jointly created as a side-effect of social tagging, into Semantic Web ontologies [90]. The advantages of this new form of ontology engineering are striking. First, the content captured by the ontology is in fact the result of a true collaborative process; the ontology is a "shared conceptualization" and it is kept up-to-date by the user community. Second, by moving from a loosely structured folksonomy to an ontology, the underlying application can leverage all benefits of semantic technologies with respect to information and data management.

On the other side, Semantic Web application and tool designers started adopting core technologies and practices from the Web 2.0 world in order to enable a greater interaction and community building. From the beginning, the ground assumption in this context was that by doing so, the Semantic Web will be able to create similar viral effects as successful instances of Web 2.0 systems, which enjoy amazing popularity and growth. A second rationale was strictly related to development and maintenance costs. Thinking back at the experiences in building knowledge bases a decade ago, which proved to be not only tedious and slow, but extremely costly and impossible to maintain, the perspective of replacing this traditional work model with a more community-oriented approach looked understandably promising. A good example for these developments are semantic wikis [120, 134]. Just as content is seamlessly created and updated collaboratively in a conventional wiki, semantic wikis alter the existing wiki syntax with means to define and refer to classes, instances, and properties in a Semantic Web knowledge base. Similar ideas can be found in semantic blogs and tagging systems.

There are several approaches bringing together Web 2.0 and ontology engineering. Braun and colleagues proposed an ontology maturing process based on Web 2.0 concepts consisting of four steps: emergence of ideas, consolidation in communities, formalization, and axiomatization [17] . They regard the evolution of an ontology as the community-driven process of maturing from tags to more formal knowledge structures. OntoGame[6] takes a different path to stimulating user participation, by designing a series of games that hides tasks related to semantic content creation behind entertaining multi-player games (see Chapter 5). The motivations behind the Linked Open Data initiative or related projects such as DBpedia[7] are very much in line with the ones that are assumed to have led to the success of Wikipedia: community sense, reciprocity, reputation, and altruism [75]. These values are fostered by the people involved in this endeavor, but also by the underlying principles and technologies, which promote openness, transparency, collaboration, and ideas exchange.

1.4.2   THE ONTOTUBE VIDEO ANNOTATION GAME

According to the Wall Street Journal [42], YouTube hosted already more than 6 million videos back in 2006, and the total time people spent watching YouTube videos within its first year summed up to 9,305 years. The amount of metadata available for feasibly managing this vast multimedia

---

[6]http://ontogame.sti2.at/
[7]http://dbpedia.org/

collection is, by comparison, very limited. At the same time, the benefits of having such metadata, in particular as rich semantic annotations, could be significant. The availability of Semantic Web descriptions using globally known identifiers and entities would enable the discovery of novel links and connections within related content within and outside YouTube, which in turn are the core building block of intelligent search and recommendation feature. Starting from these promising prospects, the OntoTube[8] casual game tries to turn the task of semantic annotation of videos into an enjoying, even addictive user experience—in response to the actual costs that this annotation exercise bears in a more traditional, paid-labor scenario.

The game is conceived as a multi-player game. The players are shown a randomly chosen YouTube video, which starts playing immediately, but can be stopped or fast forwarded at any moment. For each video, the players have to agree on answers for a set of questions derived from a background video ontology. The more questions the players manage to answer consensually, the more points they earn. The number of points depends on the difficulty of the question, i.e., we assume that indicating whether a video is in black/white or color is easier than achieving consensus on the general topic of a video. Figures 1.2 and 1.3 show two screenshots of this prototype. The first one shows one of the introductory questions, which determines the basic type of the movie. The second is an example of an advanced question, where we try to obtain further information about the genre. Siorpaes [113] makes an analysis of the effects the high number of choices in this second instance have on the game experience. An alternative could be to apply automatic techniques to try to restrict the number of genre candidates, using human computation only to validate the results of the algorithms; the downside of this approach is that it required techniques with high precision—in order to ensure that most of the time the correct answer is actually presented to the players.

The appeal of the game is in the intellectual challenge and the social experience it creates. Players must quickly grasp the content of a video. Giving correct and consensual answers to the various types of questions asked in the game can be very challenging as each game round takes only two minutes. On top of this, the video controls (play, fast forwarding) are individually configured for each player. Some of the questions are relatively easy to answer, which guarantees that users have a feeling of success. Other questions are more difficult, especially when players have to enter free text, always striving to achieve consensus with their partner. There are several approaches for determining consensus when it comes to free text inputs. The simplest approach is allowing players multiple guesses, but this may prove impossible to solve if a wide range of answers is theoretically possible, thus frustrating the players [132]. A second alternative is to constrain the number of allowed answers, provided the selection of the correct choice is unambiguous for the players. As a last possibility, one could change the actual type of the game from selection-agreement to input-agreement, just as it was done in the TagATune game [77]. In this variant, the players would be shown two videos and they have to describe them using their own words. The goal is rather to guess whether they are watching the same material, rather than agree on the terms to describe it. In

---

[8]http://ontogame.sti2.at/

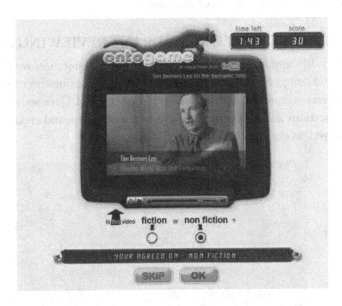

Figure 1.2: OntoTube: Determining the type of video.

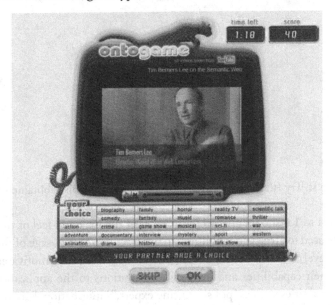

Figure 1.3: OntoTube: Choosing the video genre.

Chapter 5 we will elaborate on different kinds of game elements and discuss how they influence the behavior of the players and the results of the game.

1.4.3   THE TASTE IT! TRY IT! RESTAURANT REVIEWING APPLICATION

The Taste It! Try It![9] application is a designed semantic annotation tool for mobile devices. It collects reviews and ratings for restaurants and other types of locations for recommendation purposes (Figure 1.4). It uses semantic technologies, including Linked Data sources such as DBpedia to assign URIs to locations and other application-relevant concepts, and creates semantic descriptions of restaurants as well as star-ratings and free-text comments.

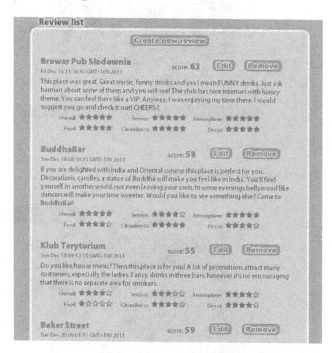

Figure 1.4: Taste It! Try It!: Reviews entered by a user and the points obtained for each contribution.

Figures 1.4 to 1.6 should give an impression of the look and feel of the game and of the core design elements used to attract contributors. In order to ensure the usage of the application as well as the appropriate level of details within the provided reviews, users are motivated by badges as a visible recognition of their capabilities and level of contributions to the application (Figure 1.5). These badges have special titles such as "polish-cuisine expert," "international-food expert," depending on the types of reviews and other activities performed by each user. The titles and other achievements can be shared within the social network of the user via Facebook, with which the application is

[9]www.facebook.com/TasteItTryIt

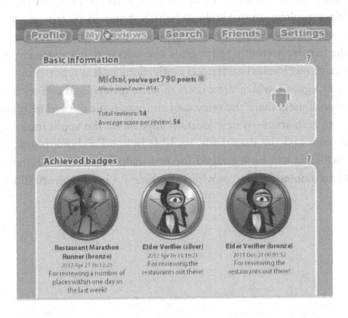

Figure 1.5: Taste It! Try It!: Badges and titles obtained by a user.

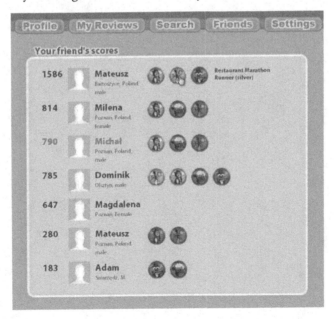

Figure 1.6: Taste It! Try It!: Comparing performance within the social network.

integrated. In addition, the users may compare their performance with their friends by checking the leader board of the application in Facebook, which shows an excerpt of the overall ranking which is close to the user level of activity—this local view is expected to stimulate competition instead of discouraging the user by presenting him goals and targets that are perceived as unattainable [1.6]. This system of game mechanics builds upon motivations such as reputation, competition, reciprocity, and conformity to a group, which drive users to write reviews and rate restaurants they visit.

If a number of users mark the same spot and assign the same category of place to it—like the name of the given restaurant—a new location appears on the application map. In addition, based on the user behavior and the data made available by the Facebook portal, the application creates a user profile, which is can be leveraged for personalization. The annotations are taken up by a semantic-based recommender system while searching for restaurants fulfilling certain criteria.

CHAPTER 2

Fundamentals of Motivation and Incentives

2.1 INTRODUCTION

In the last few years, both private and public institutions tend to involve clients and web users in product development and customer innovation processes often using various Web 2.0 technologies. Thus, they develop Web 2.0 platforms or crowdsourcing initiatives to sustain a communities of users, get new ideas from consumers, improve brand awareness, and also to outsource the workforce, reducing labor costs and improving the efficiency of content creation processes.

In this evolving domain, companies need the proactive participation of users who have to provide good quality input. Their motivations and willingness to contribute are therefore fundamental elements that companies and institutions should take into account for the implementation of the above mentioned initiatives.

Researchers from both computer and social sciences have investigated users' motivations and human willingness to participate in some well-known Web 2.0 and crowdsourcing applications. These studies reveal that people's motivations are heterogeneous and strongly connected to the uniqueness of each project [22, 23, 29, 51, 76, 80, 126, 139, 140, 141, 142, 143]. In some cases people participate because of a combination of financial compensation and learning challenges, or because they have fun, want to fruitfully spend their time, build a network of friends, or develop new opportunities for the future.

In order to shed some light on these topics, the concept of motivation and the antecedent factors that may influence the willingness of participants to contribute will be described.

2.2 DEFINING MOTIVATION

According to the Oxford dictionary, a motivation is a reason for acting or behaving in a particular way; namely, it is a desire, an enthusiasm, or willingness to do something. The concept of motivation was coined in the late 19th century but it derives from late Latin "motivus" or "movere" which means "to move," push, or influence in order to proceed in the fulfillment of a want.

In economics and psychology, motivation is considered as power that strengthens, gives direction, and triggers the tendency to continue a specific behavior [2, 10]. In other words, motivation is a physiological or psychological want aimed at satisfying an unfulfilled need. It is an internal mental state affecting the initiation, the direction (the choice of activities people make in expending effort), the intensity (the level of effort people exert), the persistence (the persistence of effort people keep over time), and the termination of a behavior.

2.3 THE CONCEPT OF MOTIVATION IN ORGANIZATIONAL STUDIES

Motivation is an internal mental state that is influenced by intrinsic desire or incentives. We can talk about intrinsic motivation if the performer enjoys the act of performing the task per se. In all other cases, a set of extrinsic incentives can be provided in order to spur individuals to perform in a certain way. In management and organizational studies, incentives are a set of instruments (e.g., money, reputation, rewards, prices, credit points, and medals) assigned by an external actor who typically evaluates the effort exercised by the performer.

One of the first examples we find in organizational literature is the concept of motivation elaborated by Frederick Winslow Taylor in Scientific Management theory. He argued that workers do not naturally enjoy their work so they should be motivated by monetary payments. During the 1930s, Elton Mayo (1880–1949) was involved in a long term experiment at the Hawthorne factory of the Western Electric Company in Chicago, and concluded that workers could be better motivated by having their social needs met while at work. Mayo introduced the Human Relation School of thought which focuses on the effects of communication between managers and workers, the manager's involvement in working lives, and the constitution of working in groups on workers' motivations.

Later, many other researchers studied motivations from various perspectives and identified various theories that can be classified as content and process theories. Content theories deal mainly with needs and what motivates people to fulfill these needs. For instance, Maslow introduced the very famous hierarchy of needs; Alderfer proposed a more simple pyramid composed by existence, relatedness, and growth needs; Herzberg analyzed motivations and hygiene factors; and finally McCelland proposed his achievement need theory. Process theories deal with the "process" of motivation, namely the way in which expectations, desires, and needs spur people to behave in a certain way. For instance, Vroom introduced the expectancy theory in 1964 arguing that motivation is strongly correlated to the individual's expectation that the behavior will lead to a particular outcome. Locke proposed the goal setting theory demonstrating that the setting of a goal (the challenging and clarity levels, and the periodicity of performance feedbacks) has impacts on motivations and performances. Adams developed the equity theory suggesting that individuals feel satisfied if they perceive their rewards to be equitable, i.e., fair, in comparison with those received by others in similar positions.

In the last few years many other studies have focused on motivation and a rich, multidisciplinary literature has contributed to the definition of an extremely lengthy list of motivation factors. As a consequence, in the next section we propose to use a simplified set of the most prominent elements that can be relevant for semantic content creation tasks.

2.4   RELEVANT VARIABLES FOR SEMANTIC CONTENT CREATION TASKS

First of all we need to carefully analyze the relevant task-related variables that might affect the performance of the users of semantic content creation tools. Broadly speaking, these are related to individual user characteristics (e.g., ability, motivation) and social, technological, and environmental/contextual constraints.

2.4.1   THE GOAL OF SEMANTIC CONTENT CREATION

People handle their activity toward specific goals which are terminated by aspiration, satisfying, impatience, and discouragement mechanisms. If the goal itself is interesting enough, the individual's motivation will be high. In many cases, a challenging goal might address the interest of individuals and catalyze interests and motivation. Specificity and reasonable levels of difficulty are also considered as important elements related to the amount of effort and persistence that individuals will use to achieve their goals (see [81]).

There are also positive relationships that will help us better understand the goal and its influence on performance.

- There is a positive relationship between the difficulty of a goal and the expected performance. Only when the goal is considered too difficult or impossible to achieve does frustration take over and the goal is abandoned by individuals. See Figure 2.1.

Figure 2.1: The relationship between performance and difficulty of a goal [81].

- There is a positive relationship between the clearness of goals and an expected performance. Clear and specific goals tend to produce better results than general and undefined ones. In an uncertain environment in which goals cannot be clearly specified at the beginning, feedback and collaboration help individuals to clarify and agree on the meaning of the goal.

- There is a positive relationship between the level of participation and the level of goal acceptance and congruence. Participation in goal-setting is a strong key to make people accept the goals (and make them clearer).

2.4.2 THE TASKS

The task, or more typically an ordered collection of tasks, is a set of actions into which a job, the annotation or content creation activities, can be divided. Internal motivation of individuals can be affected by the nature of the task which refers to both job dimensions and the relation between the job and the performers.

- The skill variety is the degree to which a job requires a variety of activities in carrying out the work and involves the use of several individual skills. Competencies are positively related to the ability to perform a variety of activities.

- The task identity is one factor of the job that influences the experienced meaningfulness of work. According to Griffin [48], task identity is the degree to which the job requires completion of a whole and identifiable piece of work that is done from the beginning to the end with a visible outcome [51].

- The job autonomy and discretion are positively related to the desires to self-organize experience and behavior according to the individual's integrated sense of self, the experience of integration and freedom, and the essential aspect of healthy human functioning [29].

- The level of tasks specificity/clearness is positively related to performance; when a task is not well defined participation might help individuals to understand it.

- The degree of amusement individuals feel while carrying out the task.

2.4.3 THE SOCIAL STRUCTURE

The social structure is the enduring and relatively stable pattern of relationships among various agents or groups of agents. It can also be considered as a set of institutionalized norms, cognitive frameworks, and cultural patterns that define and address actions of agents in a social system [82]. This pattern of relationships is wrought according to the set of norms, communication channels, coordination mechanisms, beliefs and views, feelings, etc., of people of varying social positions within a group, a firm, or any other social system. In traditional and positive studies, social structure is usually considered as a set of situational constraints or barriers, which may be beyond the control of one individual or a group. For instance, they may be tools, equipment, procedures, internal climate, relationships, information, values, and knowledge commonly shared by the individuals which influence the choices, opportunities, and/or challenges that individuals possess and foresee [40, 46].

The social structure of business companies are characterized by a variety of hierarchical levels in which the system of relations strongly depends on users' power and reputation. The latter is usually based on the supervisor's feedback. In this case, the role of supervisor (principal) is quite strategic. He/she might clearly communicate expectations, provide the necessary training, or clarify the goals assigned by the management; he/she can remove or minimize constraints, recognize positive behaviors, or reward people (agents) for their performance. In this context, relationships between principal(s) and agent(s) are not neutral, and two main issues should be taken into account: the principal-agent problem and stewardship.

- The principal-agent problem occurs when cooperating parties have different goals and visions of labor [62, 106]. Specifically, agency theory is directed at the ubiquitous agency relationship in which one party (the principal) delegates work to another (the agent) who performs that work. This is the typical case of the employer/employee(s) relationship in which a principal hires an agent to pursue a specific interest. The two agents may not share the same interests, and a situation of information asymmetry, uncertainty, and risk may occur. Considering the contract between parties as the unit of analysis, the focus of the theory of agency is on determining the most efficient contract given assumptions about people (e.g., self-interest, bounded rationality, risk aversion), organizations (e.g., goal conflict among members) and information (e.g., information is a commodity which can be purchased). The solution to this problem, and closely related to the moral hazard problem, is to ensure the provision of appropriate incentives which spur agents to act in the way the principal requires. Various mechanisms can be used as incentive mechanisms such as piece rates, (share) options, discretionary bonuses, promotions, profit sharing, efficiency wages, deferred compensation, and so on [103]. In the case of tagging and ontology population, in a hierarchical context, the systems of incentives can be based on performance-related payments (which might depend on the quantity and the quality of content annotation/population provided) or profit sharing [63, 73]. Some studies [31] show that when agents are placed on individual pay-for performance schemes or performance evaluations, they are less likely to help their co-workers, and free-riding effects increase. These negative effects are particularly important in those jobs (such as the ontology creation) that require working in teams to develop a shared conceptualization [5].

- In an organization context, stewardship refers to management's responsibility to correctly utilize and develop its resources, including its people, its property, and its financial assets. It also refers to thanking and recognizing donors. This includes organizing thank you phone calls, recognition events, and conveying the impact that the donor's gift has had.

In a community of peers, hierarchy is considered neutral, and relations among agents do not heavily depend on the role individuals have in the firm. To regulate the quality and the quantity of user contributions and ensure a sustainable level of user participation in the community, it is important to adapt the individuals' and groups' rewards to the community's social structure, such as the needs of community members, their style of contributions, their past performance and practices, their system of values, etc. In particular, in communities, situations involving interactions with other human beings are characteristically more heavily laden with emotions than any other situations [111].

Also, the majority of human emotions are social in nature and specifically developed so as to make the active group cooperation feasible [127]. Focusing on how behaviors change through motivations the following elements should be analyzed: social identity, feedback processes, justice and equity, trust, altruism, and reciprocity.

- The notion of social identity [119] is defined as an umbrella term used throughout the social sciences to describe the way individuals label themselves as members of particular groups (social groups, communities, nations, classes, etc.). Categorizing themselves into groups provides individuals with a sense of belonging that contributes to a positive self-concept. The process of interaction among peers also develops a consistent set of behaviors that reinforce the identity of the person. Theoretical analysis as well as the review of empirical studies of the relationship of social identity with motivations and performance leads to the conclusion that identification is positively related to work motivation and task performance.

- Positive and negative feedback focus on how behaviors change according to positive and negative reinforcement or punishment, and it concentrates on the relation between past and future behaviors. Individuals answer on impulse when choosing from several possible responses. Usually, the responses they choose are the ones that in the past obtained positive results. For example, an individual will repeat the action which in the past he/she did correctly and obtain a prize. In those cases, awards and punishments by peers can be used to reinforce a favorable behavior or stop a disliked one.

- Justice and equity refer to fair treatment with respect to the effort-reward balance. The perception of equity is determined by the comparison between (i) the balance between individuals' own outcomes (payment and rewards such as salary, benefits, intangibles, recognition, etc.) and individual inputs (work and effort such as hard work, skill level, tolerance, enthusiasm) and (ii) the balance between a partner's outcomes and inputs [4, 19]. It could also be based on the so-called procedural justice: a fair treatment in terms of how decisions are made about things that affect them in the workplace.

- Trust, altruism, and reciprocity are essential to human interaction especially in dealing with too complex tasks. In many experiments it emerges that people are often altruistic, trust others, and reciprocate the benevolence of others far more than the economic models of "rational" human selfishness predict. This is the case of Wikipedia contribution in which altruism, reciprocity, community, reputation, and autonomy are considered antecedents of user motivations [75]. Trust grows when each side's contribution is reciprocated by the other's, but not if reciprocity is withheld, and trust shrinks rapidly if one party abuses the other's trust by acting opportunistically (this is the case of free rider opportunism). The free riding behaviors are dramatically reduced in small groups in which individuals count on the others' contributions.

2.4.4   THE NATURE OF THE GOOD

In economics a good is any object that increases the utility of the consumer/product directly or indirectly. In semantic content creation, the good is the outcome of the annotation or ontology creation processes. The nature of good is a stylized description, in game-theoretical terms, of the relationship between what good is produced and who consumes it. We can distinguish four types of goods: private goods, public goods, common resources, and club goods.

Private goods are both excludable and rival. Most goods in an economy are private, as for example a piece of bread. A piece of bread is excludable because you can prevent someone from eating it—you just do not give it to the person in question. A piece of bread is rival because if one person eats it, the other one cannot eat the same piece of bread.

Public goods are neither rival nor excludable, i.e., people cannot be prevented from using a public good, and one person's enjoyment of the good cannot reduce another person's enjoyment of it. Street lighting is a common example of public good. Once a country is defended from aggressors, it is impossible to exclude any person from the enjoyment of this protection. The common feature in public goods settings is the existence of external factors. In public goods problems, individuals can use private resources to provide goods that have positive externalities for others. Moreover, when one person enjoys the benefits of being protected it does not interfere with the enjoyment of any other person being protected.

Common resources are rival but not excludable. For example, fish in the ocean are rival goods. If someone catches fish there are fewer fish for others to catch. However, the fish are not excludable as it is difficult to charge fishermen for the fish they catch.

Club goods or natural monopoly are goods that are non-rival but excludable. Cable television is an example of such good. To have cable television you need to pay a fee; those who do not pay the fee cannot have access to this television. But enjoyment of the program transmitted by cable television by one person does not preclude other individuals that paid the fee from enjoying it at the same time.

The nature of good influences the structure of the appropriate incentive model in a dramatic way. In the case of private good it is possible to impose a price to be paid by the final user to take advantage of the good. Exclusivity of private good allows the possibility to check that each final user pays this price. The producer of the good in this case receives a material reward that becomes the main motivational factor for production. In the case of public good it is impossible to adhere to the price if it was imposed on the good given that it is impossible to exclude an individual who is not paying the price due to the non-excludability of the good itself. Therefore, individuals tend to consume the good without contributing to its production. This problem is referred to as a problem of free riding—enjoying the benefit of a good without paying for it.

Annotation of semantic content in many cases can be seen as an example of public good. Typically, the final beneficiaries of annotation are the same individuals that perform the annotation task. It is impossible, and meaningless, to exclude any user of the portal from benefits deriving from the semantic content annotation once it is done. At the same time an annotation task requires attention and time from the user and it is a costly activity. There is a strong individual inducement to avoid undertaking the activity of annotation, therefore avoiding the individual cost of annotation, but to take advantage of the benefits provided by annotation activity carried out by other users. Therefore, the main concern of the designer in the case of public goods is dealing with the free rider problem and inducing all users to contribute.

Annotation of semantic content becomes a club good when only the users that annotate benefit from it. In this case, designers should concentrate on the involvement of users in the club and also stress the level of members' contribution.

Annotation of semantic content can be seen as a private good when users that annotate are different from the ones that benefit from it. In this case, incentive mechanisms should focus on motivating the third party to perform the task in a required way (i.e., principal-agent problem).

Public goods have been extensively studied in experimental economics, and many experiments have been conducted to test strategies on increasing contributions in public good (although a little dated, Ledyard [78] provides an excellent survey of experimental literature on the point). From these studies we can produce the following rules:

- Pay-off structure. People improve their input if their personal return on contribution is high and the return on not contributing is low. People also improve their contribution when the benefit to others (deriving from their contribution) increases [15, 66]. Another positive effect on the level of contribution to public goods may derive from the allocation of a lump sum to be spent by the group rather then by each person [6]. While in laboratory experiments pay-offs are clear and carefully explained to subjects, benefits of semantic content creation may not be so clear to users. Therefore, designers should clarify which benefits users can get from the platform and how these benefits may improve their life/work/use of the platform. This kind of awareness can be created through the implementation of a participatory design process in which users point out their desiderata and get the designer's feedback. It is possible to introduce a system of points or badges that users get for their actions. These points/badges can then be transformed into personal benefits ranging from monetary rewards or additional pay-off in a corporate setting to the construction of a status in the community.

- Group size. Results on the effect of the group size on contribution to public goods are not unique. Numerous studies demonstrate that cooperation declines as group size increases [15, 37, 74]. In a large group it is more difficult to monitor everybody's contributions. People may expect others to contribute and think that the lack of their small contribution will not be noticed. In addition, in larger groups it is more difficult to encourage others to contribute by personal example, while defecting anonymously is a very attractive strategy. This evidence encourages the designer to create a network of smaller groups of users. This solution is feasible for corporate environments where this division is naturally accepted. However, even if it is not possible to divide users into smaller groups it is possible to obtain efficient levels of contributions. Several studies have found that decline in contributions with increase in group size vanishes quickly [37, 79]. Yamagishi and Cook [145] and Isaac et al. [61] demonstrate that contributions increase as the size of the group increases. In particular, Isaac et al. [61] study interplay between group size and pay-off structure. Experimental subjects make their choices over the Internet at their own locations. They show that groups of 40 and 100 provide the public good at higher levels than groups of 4 and 10 for lower levels of benefit from public good while the difference is non-significant for a high level of benefit. In the experiment with very low individual benefit from public good they found the same levels of contribution in large groups compared to small groups. Marwel and Oliver [89] suggest that large groups are more likely to contain a critical mass of individuals whose interests are served by providing highly non-rival good. Heterogeneity of the group members' interests and resources encourages formation of the critical mass ([41]).

• Thresholds and provision points. In the case of threshold or provision point, if enough resources are collected then the public good is provided and all receive its benefits. If too few resources are collected, then the public good is not provided and no positive externalities are enjoyed. Theoretically there exist multiple equilibriums in which the public good is provided exactly with each individual contributing a share of costs. These equilibriums differ by the share of the cost covered by each single individual. Experimental data seem to suggest that if the threshold increases, the contributions will increase but also the probability that the target will not be reached [60] will rise. As a consequence, designers should define the target the project requires and the average number of actions that each user needs to perform. This target should be feasible for members of the community and their single action should be recognized as very important to achieve the common goal.

• Efficacy. The main reason for a low level of contribution to public good is that contributors do not have the perception of how significant an effect they produce to achieve their common goal. If a person can have a noticeable effect on the outcome, the levels of contribution will increase. A way to increase this perception is to introduce thresholds. If an individual believes that the group is close to the threshold, then each single contributions may help the group to reach it. The perception of efficacy often induces people to cooperate more [68, 104].

• Group identity. In social identity theory [118] it has been studied that categorizing individuals into a group increases altruism behaviors among the group, and a higher level of contributions is expected. Besides, introducing a competition among groups in public good games with threshold may reduce free riding behaviors [104].

• Communication. Communication among group members significantly increases levels of contribution [79, 97]. Ledyard [78] assessed that the effect of communication is limited to small groups with size N smaller than 15 individuals. Dawes, McTavish, and Shaklee [26] investigated what type of communication affects the contributions. It emerged that if people discuss the topic of contribution, their commitment will rise and the contributions will double. Dawes [28] argued that the key effect of communication comes from eliciting group identity.

- Moralizing. Dawes et al. [27] showed that contributions of people exposed to a moralistic speech about group benefit, exploitation, ethics, and so on, are higher than the contributions of people that were not told anything. Although this speech may also affect also social pressure, experimental demand, emotional appeal, etc., it works at motivating people to contribute. This strategy can be easily implemented in corporate portals where the head of the company, department, or group gives a motivational speech on how the implementation of semantic ontology is important for the company and on the need for everyone's contribution.

- Public disclosure of choice versus anonymity. Making public the choice of each member of the group increases individual contributions to public goods [38]. People tend to imitate those around them. In a university fund raising campaign, Frey and Meier [39] found that information about the average contribution in the past has a significant impact on contribution. In a field experiment on MovieLens providing information on the performance of a median user significantly increased monthly movie ratings of below median users while above median users did not necessarily decrease their contributions [21]. Report on the average contribution to the community is an easy and effective instrument. Each member of the group should be able to see what other members did and how the group stands in relation to other groups.

To sum up, the designer has a wide range of tools to influence the behavior of users of the platform, and in the next section a tool for domain analysis is provided.

2.5 THE FRAMEWORK

Summarizing these variables, Figure 2.2 provides the framework a designer can use to examine motivational sources for semantic content initiatives and to define a set of incentives that might spur individuals to act according to his/her aims.

Essentially, we can map any situation as a specific combination of traits (i.e. goals, tasks, social structure, and nature of the good). Then, for each combination of traits, we can specify a mix of explicit incentives and motivational mechanisms that can elicit participation in the given situation.

In order to better understand how to proceed, we also developed a procedural ordering of practices that a designer can carry out in order to design the most effective set of incentives (see Figure 2.3). We do not propose merely a design process allowing for the incorporation of some received wisdom into technology; we rather propose a methodology that encompasses the ability to analyze the social nature of any work environment and conduct. Specifically the set of methods and techniques we use refers to "mechanism design" in the field of economics. This approach stresses a "value-chain" outlook on the design process, clearly distinguishing problems regarding motivation/participation in the design process on the one hand, and motivation/participation of users once the tool is in place. Mechanism design enables the analysis of the goal, the tasks, the social structure, and the nature of goods being produced prior to the design of the application and to propose the incentive structure for the application.

Goal		Tasks		Social structure	Nature of good being produced
Communication level (about the goal)	High	Variety	High	Hierarchy-neutral	Public good (nonrival, nonexclusive)
	Medium		Medium		
	Low		Low		
Participation level (in defining the goal)	High	Specificity	High		
	Medium		Medium		
	Low		Low		
Clarity level	High	Identification	High	Hierarchical	Private good (rival, exclusive)
			Low		
	Low	Required skills	Highly specific		
			Trivial		
			Common		

Figure 2.2: Framework for studying motivational antecedents.

Ideally the process of design and development of an incentivized application should start from the analysis of the concrete situation. The field analysis is crucial to identify the motivations of both individuals and the social groups to which they belong. Direct observations, interviews, and questionnaires are very effective techniques that can be used to unveil and better define the crucial elements. The raw knowledge is then analyzed when modeling the situation by using game theoretical predictions about the behaviors of the actors described in the model. Given a set of goals, and the results of previous experiments already described in previous paragraphs, this model enables the analysts to design a set of incentive schemes that would spur users to behave in line with the desired outcomes. The creation of the prototype, which should be the simplest possible solution that can effectively support the users, follows. The resulting prototype is first tested in a controlled environment such as a laboratory. The laboratory allows the experimenter to test a hypothesis with artificially controlled conditions, manipulating or eliminating extraneous factors. As soon as the previous hypotheses are confirmed, a sequence of experiments can be organized to fine tune the

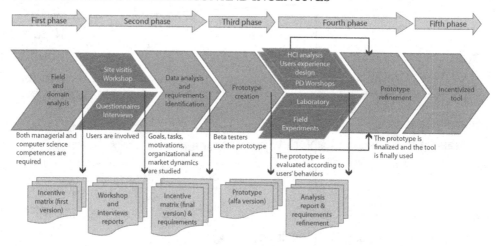

Figure 2.3: The ideal process of design, development, and evaluation of an incentivized application.

set of incentives that are embedded in the tool. The design of each experiment may depend on the results of previous ones. In the last phase—the field experiments—the tool is tested in the field while mimicking the situation in the lab. This fine tuning process should increase the complexity of the trial since the experiment gets closer to reality. For example, adding new realistic components such as real actors (i.e., community members), tasks (a daily activity that actors usually carry out), and situations (the field and the social structure which actors belong to). As the level of control over the ability to manipulate the tasks, the goals, and the inner motivations of individuals diminishes the experiments gain the ability to adequately simulate real and concrete environments/situations. This might continuously require new changes to the tool that is finally introduced in the field. This process is iteratively repeated until the tool is finalized.

In the following chapters we will use this procedural ordering of practices and Figure 2.2 to analyze the various case studies and identify incentive structures and features to implement.

CHAPTER 3

Case Study: Motivating Employees to Annotate Content

3.1 AIMS AND OBJECTIVES

Telefónica Investigación y Desarrollo (TID) is an international company of more than 1,200 employees with offices in Spain (Madrid, Barcelona, Boecillo, Granada, and Huesca) as well as in Mexico (Mexico D.F.) and Brazil (Sao Paulo). As a typical R&D company, most of the TID staff are highly skilled telecommunication and informatics engineers, physicists, and mathematicians but also business administration experts. The staff is hierarchically organized into three main work areas (management, technological, and scientific divisions) and at five levels (directors, managers, heads of departments, project managers, and R&D engineers). All of them have access to an internal corporate portal which is one of the main corporate communication tools (blogs, forums, wikis, document and knowledge management tools, etc.) provided by the company. The enormous amount of information and services provided by the TID intranet portal has caused a noticeable fall in its efficiency and dynamism as a repository and a collaborative environment. In particular, it is difficult to gain access to the right information asset when it is needed.

To solve this problem, the OKenterprise prototype has been developed with the aim of enriching the corporate portal content with automatic and user-guided semantic annotations. These annotations can later be used to improve the performances of the portal by providing semantic searches and semantically related content recommendations.

In this chapter, we present the process we followed in the development, testing, and validation of the OKenterprise prototype led by Telefónica Investigación y Desarrollo (TID). In particular, we (i) observed the participants and users interviews, (ii) organized and ran a laboratory experiment with a limited number of users, and (iii) rolled out the prototype involving the TID employees.

3.2 METHODS USED

In order to evaluate and provide design recommendations for an incentivized prototype, we adopted the procedural ordering of practices described in Figure 2.2. Following the phases proposed in the Chapter 2, we first analyzed the domain and the state of the art of technologies used to manage knowledge in large companies. Then, in the second phase, we focused our analysis on the employees' motivations and the motivations of social groups (e.g., team work, communities of experts, etc.). In order to unveil and better define the crucial lines of behavior and desiderata of TID employees, we conducted direct observations, interviews, and a workshop (these latter are based on the focus group methodology). In the third phase the prototype was developed in a very simple version but usable by final users. In the fourth phase the resulting prototype was tested in various scenarios. We conducted our analysis in a controlled environment, namely a purposely set laboratory. As soon as the first hypotheses were confirmed and the prototype refined, a sequence of experiments and participatory design workshops were organized to fine tune both OKenterprise and the set of incentives that the tool embedded. This fine tuning process typically increased in complexity as the experiments got closer to conditions which carefully mimicked reality (field experiments). For example, this was done by progressively adding new realistic components such as real actors (e.g., community members), tasks (a daily activity that actors usually carry out), and situations (the field and the social structure to which actors belong). This way the process continuously involved new changes to the tool which was eventually introduced into the company. In the fifth phase the tool was finalized and released.

In detail we used the following techniques:

Semi-structured interviews are flexible interviews in which the interviewer, following a general framework of themes to be explored, asks questions related to what the interviewee says. This adjusting process helps interviewers to tailor questions to the interviewee needs and context/situation. Semi-structured interviews were conducted to obtain knowledge about the general context of the platform and to understand the users' expectations, to measure the users' satisfaction, and thus to identify further requirements.

Requirement workshops are structured meetings where a selected group of stakeholders works together to define or refine a set of project requirements. During these workshops, participants discovered, defined, and prioritized the requirements that were needed for the OKenterprise.

Usability tests are techniques used in user-centered interaction design to evaluate a prototype by testing it on users. We focused on the following usability dimensions: effectiveness, efficiency, and satisfaction.

Laboratory and field experiments are tests or investigations carried out in a laboratory or in the field. The laboratory is an artificial environment which allows the experimenter to test hypotheses with totally controlled conditions, manipulating or eliminating extraneous factors. Field experiments have the advantage that outcomes are observed in a natural setting rather than in a contrived laboratory environment. For this reason, in social science, field experiments are sometimes seen as having higher external validity than laboratory experiments.

3.3   CASE STUDY DESCRIPTION: THE OKENTERPRISE

OKenterprise is an extension for Google Chrome and its logo appears on the right-hand side of the Google Chrome address bar. The position of the OKenterprise logo is relative to the window so that when the page is scrolled the icon stays accessible. Clicking on the added OKenterprise logo, the OKenterprise side-bar expands as depicted in Figure 3.1.

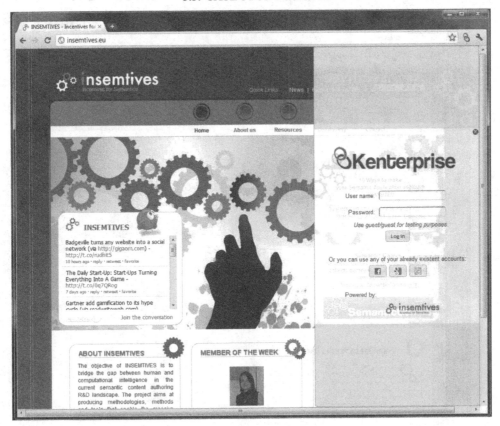

Figure 3.1: The OKenterprise side-bar showing the login page.

The side-bar shows the authentication page which allows the user to log in using four distinct authentication schemes: Facebook, Google, Twitter, and OKenterprise accounts. In the case of the TID Internet corporate portal the accounts are synchronized with the login and passwords used in the portal where the semantic annotation capabilities are being used.

Once the user is identified, the OKenterprise side-bar appears with updated content. As depicted in Figure 3.2 it contains six modules: Home, Annotate, Navigate, Search, Configure, and Help. Users can personalize the side-bar, collapsing some modules or changing their order by dragging each module to the desired location.

The Home module is the first and includes four sections which are:

1. User: shows some information about the current user.

2. My knowledge cloud: displays the most added annotations. The size of annotations shows how many times the users added them. The cloud is also visible to the rest of the community. Since

Figure 3.2: The OKenterprise side-bar after user authentication.

each user annotates topics in which he/she is interested, the knowledge cloud may represent the areas of expertise of the users.

3. My badges: includes the badges the current user has unlocked (shown in colors) and the ones he/she can still unlock (shown in black and white). To unlock new badges, users have to interact with the tool; for example, annotating new content.

4. Week Top Contributors: provides the user with a micro-leader board regarding the most active users of the current week.

The Annotate module includes three sections which let users annotate the content exposed in a web page. Users can select the web page or any part of it, add one or more annotations or semantic annotations, check his/her annotation cloud reassigning a meaning to an annotation as well as assigning a meaning to a free-text tag, and finally submit the annotations (see Figure 3.3 and Figure 3.4).

Table 3.1: The user interface evaluation of the OKenterprise prototype

Scenario	Problem	Recommendation
	1. Two text boxes: Because of the layout of the text boxes are quiet small. Therefore, the visualization becomes really difficult because of lacking space. **2. Not understandable symbols in the suggestions:** The colon behind the suggestion is irritating as long as there is no concept matching the word. **3. Possible dispensability of definition:** The definition covers parts of the results and thereby hinders the user to compare all results. Sometimes the overlying concept is perfectly fine for creating an understanding of the tag.	1. Only offer one text box or layout the text boxes vertically to get the full wide for displaying the results. 2. Avoid textual descriptions which may be too formal for the intended audience of OKenterprise. 3. In some cases the tool tip is not necessary to explain the meaning. A symbol (e.g., "?") should be provided which can be clicked if the user desires to see the definition.
	4. Too many buttons: Suggestions should not be triggered by the user explicitly, but rather be performed automatically in the background. For instance, if the button "Submit annotation" is deleted, submission con be done with the enter button of the keyboard. **5. Differentiate suggested confirmed and unconfirmed tags:** If there are too many suggestions the user is not able to distinguish between them and own tags. **6. Save state is unclear:** The process of submission is too complex.	4. By offering too many options the visible area is overloaded with too many buttons. 5. Functionalities related to suggestions and loading should be undertaken in the background. To avoid irritating the user with own and suggested tags, there should be a non-intrusive area for suggestions. 6. Submitting annotations should be carried out automatically at the end of the annotation process or when the application is closed, if a tag is added.
	7. Redundant information in the tool tip: The concept is the tag itself with further information in brackets. **8. Visualization that no concept is bootstrapped is missing:** The user needs to hover over a tag to become aware of the fact that it is possible to enhance free text with concepts.	7. The tool tip should provide only additional information. If the term for the concept is similar to the tag itself, just show the information in the brackets. 8. Use the additional free space to show the beginning of the description. An idea would be to bootstrap the description to get keywords for the meaning of the concept.

The Navigate module includes two sections which let users obtain content recommendations related to the web page they are currently visiting. These content recommendations are based on the annotations made on the content the current web page comprises.

The Search module enables users to look for content or/for other users. Using auto completion features the users can identify some semantic keywords and search for related documents or parts of documents.

The Configure module enables users to customize the tool to their needs, and the Help module provides a user guide of the tool.

The OKenterprise source code is available in the INSEMTIVES SourceForge repository at `http://sourceforge.net/projects/insemtives/`.

According to the procedural ordering of practices described in Figure 2.2 we organized our analysis as follows.

3.3.1   FIRST AND SECOND PHASES

In the first and second phases of our analysis, we collected data on people's motives and motivation drivers by interviewing 11 employees of TID as representing the community at large. Each semi-

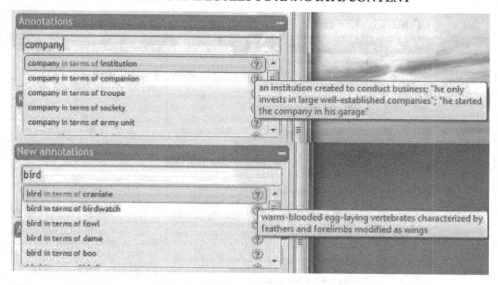

Figure 3.3: "New annotations module" including the auto-completion feature.

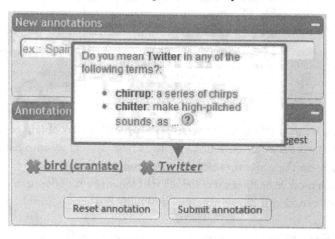

Figure 3.4: The "annotation cloud" module.

structured interview was conducted by two interviewers, took 60 to 90 minutes and was recorded on audio tape. These recordings have been transcribed and analyzed descriptively according to ex-post categories. Additionally, a focus group discussion with six TID employees was conducted, focusing on usage problems of the existing system and on for innovative solutions that might overcome these problems. We analyzed the interviews and classified the TID portal features and we also depicted some motivational antecedents as can be seen in Figure 3.5. This figure is a revised version of the Figure 2.2 described the Chapter 2. In detail:

Goal		Tasks		Social structure	Nature of good being produced
Communication level (about the goal)	High	Variety	High	Hierarchy-neutral	Public good (nonrival, nonexclusive)
	Medium		Medium		
	Low		Low		
Participation level (in defining the goal)	High	Specificity	High		
	Medium		Medium		
	Low		Low		
Clarity level	High	Identification	High	Hierarchical	Private good (rival, exclusive)
			Low		
	Low	Required skills	Highly specific		
			Trivial		
			Common		

Figure 3.5: Motivational antecedents in the case of TID.

- The goal of the annotation tool is to improve the search and navigation experience in a corporate knowledge base. TID providers of annotations might have two different motivations to annotate. Users—by annotating resources—can show their areas of competence and interest to the community. Moreover, they improve the navigation, searching, and syndicating capabilities of the enterprise portal by using annotations. The goal can be very clear and refer mainly to annotating content to improve the knowledge portal services, but it can also deal with individual goals such as being recognized as an expert in specific topics (described by the knowledge cloud of the tool).

- The task is a typical annotation task. This means that it is very repetitive, with a low level of variety and a high level of specificity, and lacks a fun element. The required skills of the agents to complete the annotation task are very basic, but the task requires a high level of knowledge to understand and effectively annotate the content.

- The social structure is quite complex and various dynamics co-exist. Employees work in teams and communities of experts but they also work in the company which has a strong hierarchical structure. Visibility, reputation, career development, and money are all part of the mix of motivations driving the behavior of employees.

From the interviews, the workshop and Figure 3.5, the following motivation levers emerged as the ones that developers should take into account while designing the OKenterprise annotation tool: reputation, competition, conformity to a group, usefulness, altruism, reciprocity, self-esteem, learning opportunities, fun, and personal enjoyment.

3.3.2   THIRD PHASE

In the third phase TID developed a first prototype of the OKenterprise annotation tool. It supports the creation of a semantic layer on top of the current version of the TID Intranet corporate portal. Since this portal is a live portal used 24/7 on a regular basis by the TID employees, a major require-ment of this case was to include this semantic layer in a non-intrusive way, namely not affecting the current users of the portal. To accomplish this requirement the OKenterprise annotation tool was implemented and released as a Google Chrome extension.

The first version of the OKenterprise prototype allows the users of the TID Intranet corporate portal to select and annotate any textual or graphical (i.e. image) content exposed on the portal's web pages.

3.3.3   FOURTH PHASE: PRELIMINARY RESULTS

In the fourth phase we performed two different types of analysis. The first was focused on the human–computer interface and the second was based on laboratory and field experiments. In Figure 3.6, the tool is analyzed and various problems are unveiled. We listed the recommendations necessary to correct the corresponding issues. Although the OKenterprise annotation tool is focused on TID employees or, more precisely, users of the TID Intranet corporate portal, the analysis that is provided refers to general end-users.

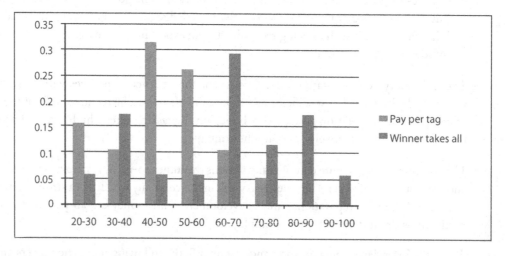

Figure 3.6: Distributions of annotations.

The recommendations described in Figure 3.6 were taken into consideration by a group of developers in TID. They met and discussed each recommendation followed by planning some changes in the OKenterprise annotation tool or even rejecting some of the proposals.

3.3.4   FOURTH PHASE: THE FIRST LABORATORY EXPERIMENT

In the evaluation of the first prototype we also ran a laboratory experiment with the goal of identifying a more appropriate set of incentives to spur people to annotate resources. The aim of the experiments was to compare the performance of two systems of monetary incentives. The first is commonly known as "pay-per-click." With this system users receive small monetary rewards for each annotation they create. The second system of monetary incentives is referred to as "the-winner-takes-all." The experiment was conducted at the Computable and Experimental Economics Laboratory (CEEL) of the University of Trento. Thirty-six university students with basic knowledge of Web technologies and platforms were randomly assigned to one of the two treatments, resulting in 19 subjects in the "pay-per-tag" (PPT) scenario and 17 subjects in the "winner-takes-all" (WTA) scenario. Subjects in both treatments were requested to tag 40 images randomly selected via Google. Subjects had eight minutes to perform the task. The results are summarized in Table 3.2.

Table 3.2: Results of the first experiment		
	PPT Model	WTA Model
Total amount of tags	901	1067
Average number of tags for each student	47.42	62.76
Maximum number of tags per student	78	96
Annotated images	38	38
Average images annotated by single student	11	11
Average tags per image	3	5
Cost for each single tag (in Euros)	0.030	0.018
Total budget of the experiment session (in Euros)	27.03	20

Figure 3.6 shows the distribution of annotations under the two treatments. We notice that the distributions differ a lot for the two payment models. In the PPT model the number of annotations is lower than in the WTA model. Thus, we can argue that the group incentivized with the PPT model coordinates on a lower level of productivity than the group of individuals who act according the WTA model.

Looking at the quality of tags users provided, we could not claim that it is lower, but we can argue that, in the WTA, the tags are less general and describe the more evident characteristics of the images such as colors and other general objects depicted in the picture. Only a few phrases were provided to describe a picture. Despite the biases that affect the experiment results, it is clear that the winner-takes-all model is dramatically more effective than the pay-per-click. These results constitute a baseline for the implementation of the tool within the TID corporate environment. However, we should keep in mind that the distance between the contextual conditions of the lab and those present at TID is still very large.

3.3.5   FOURTH PHASE: THE GAMIFICATION OF THE TASK

The prototype refinement strongly emphasized the gaming character of the annotation process. Users are provided with a system of badges that shows the achieved annotation goals (Figure 3.7).

Figure 3.7: Badges in the OKenterprise annotation tool.

Users can also have access to a scoreboard (see Figure 3.8), which shows the position and the score of the contributor in relation to the other participants. This information tends to increase the competition among users. Users can also select a single concept and see who annotated it most. In that way users have a clue about who may be the expert on specific content, increasing reputation as an incentive mechanism.

As we can see from the previous results, the OKenterprise tool seems perfectly shaped for the TID employees. Even if employees are working in a very hierarchical structure, competencies and reputation in terms of expertise matter considerably. Workers act in a very competitive environment in which individuals contribute in order to obtain a direct reward, such as a salary, or indirect reward, such as reputation. In other words, the company is considered an internal labor market, and employees compete to obtain a better position in the near future. For this reason we believe that the Human Resources department should be involved in the definition of rewards, badges, and in the measurement of the annotation activity.

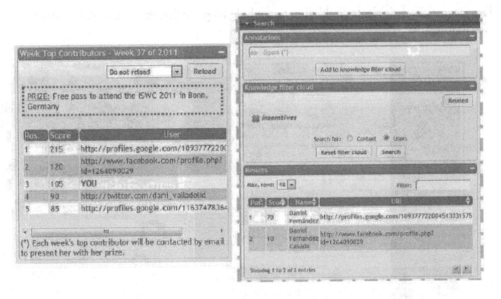

Figure 3.8: Scoreboard in the OKenterprise annotation tool.

3.3.6   FOURTH PHASE: THE SECOND LABORATORY EXPERIMENT

Experiments conducted in the initial phase of the fine-tuning suggested that competition for a prize leads to a better performance. Our next step was to define the best prize system to be adopted within TID. Apart from a monetary prize we considered another type of prize that could be easily applied within TID—social recognition. Instead of giving monetary rewards we wanted to see whether plain social recognition can help to create semantic content. Thus, in the experiment two treatments were compared: a competition for a monetary prize and a competition for social recognition.

The fact that the OKenterprise tool embeds the English version of WordNet created some challenges for the laboratory test. First of all we needed to find experimental subjects with an appropriate knowledge of English. Next, we needed to choose a domain for the experiment that would have good coverage in WordNet and would meet the English vocabulary of our prospective subjects. The most covered domains in WordNet according to our first analysis seemed to be the programming, cooking, travel, and education domains.

We chose the travel domain as vocabulary on travel should be the most diffused among students. Twenty-five university students were randomly assigned to one of the two treatments, resulting in 16 subjects in the monetary-prize competition scenario and nine subjects in the social recognition scenario. Subjects in both treatments were requested to tag the sightseeing blog created for the experiment containing information on sightseeing opportunities in some of the biggest

European cities.[1] Subjects could annotate whole blog entries, images, or individual text fragments (see Figure 3.9).

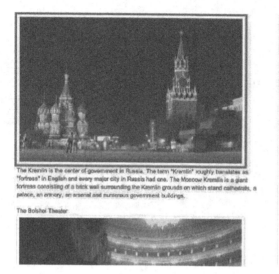 

Figure 3.9: Screenshot of the tool during the experiment.

Subjects had 15 minutes to perform the task. All subjects were paid a participation fee of 5 Euros. In the competition scenario the person with the highest score received 15 Euros. In the social-recognition treatment the person with the highest score was announced to the entire group.

At the beginning of the experiment subjects taking part in the same treatment were randomly assigned to computers. Their workstations were separated in order to prevent any interaction. A clear set of instructions for the experiments was read aloud, while subjects were following the instructions on the computer screen in front of them. Instructions explained to the subjects the task they had to perform, the payment procedure, and the time they had to perform the task. Instructions also stated that annotations would be controlled at the end of the experiment and individuals that made annotations not related to the content, i.e., attributing numbers or tags consisting of single letters or other obviously low-quality tags, would not receive any payment. The instructions were the same for both treatments except for differences in the explanation of the payment procedure.

In Figure 3.10 a few simple statistics detailing the main findings of the experiment are reported. Since we are talking about two groups of experimental subjects of different size, data are displayed as average contributions per subject. The results of the two groups are significantly different as proven by the significance level of the Wilcoxon-Mann-Whitney test. In particular, the contributions of the competitive treatment group are on average more than those of the "social" group. This means that if contributors compete for a prize (in this case a sum of money significantly higher than what

[1]http://insemtivestourism.blogspot.com/

	Competition treatment	Controls (social)
Avg. nr. of annotations	46	36
Avg. nr. of semantic annotations	31	23
Avg. nr. of free-text annotations	16	13
Maximum nr. of annotations	85	53
Pct. of semantic annotations	68%	64%
Wilcoxon-Mann-Whitney test result for total number of annotations	p-value = 0.004 [W= 119]	
Test on difference between proportions of semantic and free text annotations	Z=0.332 (p=0.37)	

Figure 3.10: A comparison between the competitive treatment and the control group motivated only by a generic appeal to the social usefulness of the annotation work

everyone else got) they will provide more annotations. Comparing semantic and free-text annotation, it emerged that there is no significant difference between the two groups. They, respectively, provided 66% and 64% of semantic annotations. In some cases, contributors provided free-text annotations that could be associated with a semantic content within the scope of WordNet but they simply decided not to use semantic. Although some of these "missed" semantic annotations can be explained away by the erratic behavior of the subjects (there are a few typos for instance or a few annotations in bad English that could not be identified correctly within the knowledge base) most of the free-text annotation cases are attempts by the subjects to provide semantic meaning by means of a complex definition (i.e., complete sentences with multiple concepts) instead of a single-concept terms (e.g., "The symbol of Berlin between Eastern and Western side" was probably caused by a knowledge gap in the dictionary as there is no "Berlin wall" concept), or because they could not find the appropriate semantic description among the ones available. One last remark regards an assessment of actions by the best performers: in both groups the leading contributors were the top semantic annotators and in the top three overall annotators even though in both cases the maximum number of annotations was supplied by subjects who tended to prefer free-text.

3.3.7   FOURTH PHASE: THE FIELD EXPERIMENT

In order to overcome the gap between the very simple lab experiments and the field study, we gradually incorporated some of the features characterizing the working context of TID by mimicking the annotation tasks required by TID employees in their day-by-day activities. The field experiment conducted at TID was aimed at understanding how a specific combination of explicit incentives and interface layout can improve the amount of annotations provided in a corporate environment. Our assumption is that people like to compete for prizes even (we might say especially) in a corporate environment. As a result, we tested the power of different prizes which may induce effort in annotations, maintaining the tournament structure of the incentive scheme (i.e., the winner takes all approach). In this first trial we introduced two kinds of prizes: (i) something that directly gives pleasure (money, a day off, a bottle of good wine, etc.); (ii) a trophy (an object that can be displayed but it does not need to have extrinsic value besides being a symbol of winning the competition and to be recognized by everyone as such).

A group of TID employees was recruited to participate in the field experiment and to use the OKenterprise annotation tool in their daily working life. The participants were 19 experts in various fields and with various representative roles. Users were asked to annotate in English, although most of the blogs are mainly in Spanish. This makes the task more complex, challenging contributors to choose the most appropriate concepts and terms. The field experiment lasted one week, from 16th January 2012, 00 : 00 to 22nd January, 23 : 59.

Users had to annotate unstructured information within the corporate website and some external blogs. For that reason, several blogs were selected, namely:

- http://www.lacofa.es

- http://www.yodigital.es

- http://blogs.creamoselfuturo.com

- https://ekiss.intranet.telefonica/comunidad/ekiss/blogs/

- http://www.rcysostenibilidad.telefonica.com/blogs/

- http://www.mujeryciencia.es/

In the field experiment the monetary prize treatment leads to the generation of a higher number of annotations compared to the social recognition treatment. Thus, in the former case participants provided 177 annotations on average compared to the 117 annotations contributed on average in the latter. The most active contributor for the first group provided 439 annotations in a week in contrast to 262 annotations provided by the best performer of the other group. Although monetary prize treatment seems to lead to higher productivity of users, the difference in the number of annotations between treatments is not statistically significant as suggested by the result of the Wilcoxon-Mann-Whitney test.

Looking at the daily contributions of the group who received monetary prizes it becomes evident that a few individuals who started the competition for the prize drove the whole number of contributions. The rest of the group contributed at a much lower level and stopped contributing in the middle of the experiment. This is a typical behavioral pattern observed in competitive environments. In the second group, participants contributed on a daily basis and the level of individual contribution did not drop as in the competition group.

We can therefore conclude that although the monetary-oriented treatment led to a higher number of annotations, only a few individuals in the group produced them. As Figure 3.11 states,

	Monetary prize	Social recognition
Number of annotations	1589	1172
Average number of annotations	177	117
Maximun number of annotations	439	262
Wilcoxon-Mann-Whitney test for total nr. of annotation	z=0,45, p=0.32	
% of semantic annotations	88,92%	71,84%
Test on difference between proportions of semantic and free text annotations	Z=11.46, p=0.0001	

Figure 3.11: The results of the field experiment.

this group provided a significantly higher number of semantic annotations; almost 89% of their annotations were semantic annotations. This difference between the proportions of semantic annotations in the two groups is highly statistically significant. It suggests that participants who were rewarded financially preferred creating semantic annotations. This comes as no surprise given that the prize was to be assigned to the person who earned the highest number of points during the competition and each semantic annotation led to 25 points compared to 5 points granted for a free-text annotation.

During the TID experiments the users generated in total 2,761 annotations distributed as follows: 2,255 (82%) annotations were semantic; 506 (18%) annotations were free text annotations. Out of the 506 free-text annotations, 111 (4% of all annotations) had one or more meanings found in the underlying knowledge base. However, it does not necessarily mean that the knowledge base included the right meaning for the free text annotation given the context in which it was implied. For the remaining 395 annotations (14% of all annotations) no meaning was found in the knowledge base.

After a manual analysis of free-text annotations, we concluded that no meaning was found for some of them because of the insufficient coverage of the underlying knowledge base (e.g., no meaning was found for "smart phone" and "social network" tags); some other free-text annotations were misspelled whereas for the correct spelling the meaning was present in the knowledge base (e.g., "doppleganger" instead of "doppelganger," "empatia" instead of "empathy"); some annotations were provided in Spanish whereas the knowledge base included meanings for the English language only (e.g., "patrocinadores" and "jornadas"); some other annotations were provided as complete sentences and not single tags and, therefore, were not assigned a single meaning from the knowledge base (e.g., "capital of Spain" and "man in the middle attack").

Focusing on the quality, we made a raw analysis of the content provided and the result is almost the same for the two groups.

In order to better understand the real motivations that drove TID employees to contribute, we asked participants to fill out a questionnaire covering the following topics:

- the level of competencies and knowledge participants had in semantics

- their participation in previous semantic based projects

- motivational factors for annotation which included the perception of the work overload

- the set of social rewards (like the visibility of the social cloud)

- the design quality of the extension

- the usefulness of the extension.

In general, users found the application very interesting and useful. In addition the performance was considered as acceptable. Participants pointed out that the material incentives did not foster too much competitiveness which would be harmful for the spirit of collaboration between colleagues. As a consequence, the inner motivations that drive people to contribute were both the prize and social reciprocity (namely they want to return a favor or they want to establish a credit with their colleagues). Much less important is the personal interest in semantics and their need to advance research on semantics. For these reasons it is mainly the prize and the badges that spurred people to actively use the annotation tool.

In the questionnaire we also asked participants to rate the level of motivation that users would have when choosing from among different prizes. The shopping voucher scored the highest, followed by a dinner in a restaurant, and one night for two people in a hotel. These three rewards were considered more valuable than time off work, recognition from the boss, and VIP access to a technology convention. These answers may derive also from the fact that the OKenterprise annotation tool was considered a game or a simple task that does not affect the career development and reputation/evaluation in the company.

The interface was considered nice looking, intuitive, and easy to use. On the contrary, the process of annotation could be improved with a more attractive list of annotated concepts. Although the results provided by the auto-completion were considered sufficient (68.75% of participants considered the auto-completion fast and useful to annotate), the utilization of domain ontologies or domain taxonomies could help users to annotate with specific and more specialized concepts. The search function could also be improved.

Referring to the auto-completion functionality, 56.25% of the participants revealed that the concept summarizations (shown in parenthesis) were not sufficient to grasp the meaning of the concept, and 50% said that the system provided too many concepts with similar meaning. Also in these cases, a domain ontology may solve the problem.

Participants also pointed out that the OKenterprise annotation tool could be very useful in their work environment and also in large communities (blogs, social networks) where text content is prevalent in order to obtain more relevant content in searches.

3.4   RESULTS AND LESSONS LEARNED

The OKenterprise tool developed by TID represents a potentially great application to incentivize semantic annotation efforts in corporate environments and, more in general, in situations where there is a need to use human work to analyze complex content. In this chapter we summarize the main results obtained from the field and laboratory tests of the tool and the related incentive schemes, outlining some of the main problems still present in the tool and for which further work is necessary. Moreover, we offer some general remarks about the problem of incentivizing annotation tools for semantic applications. As we already pointed out in Section 3.3, the testing and development cycle we employed for analyzing the OKenterprise tool identified a few robust predictors for performance and a few more effects that seem to vary under different contextual characteristics. This cycle included several rounds of experimental tests aimed at including both a more traditional testing approach for the technical features of the tool and an innovative, incremental testing of the incentive structure associated with the annotation tasks and aimed at improving performance. The method itself proved useful by helping to combine the development of both technical and incentive-related features.

Consistently—across both lab and field experiments—we found that competition between subjects helps increase the level of effort displayed. Moreover, competition seems to work both when the prize is of a monetary nature and when it is just handed out in the form of a symbolic recognition. However, there are also some clear differences: although in the laboratory the two treatment results are significantly different, the field experiment does not display statistically significant differences in the motivation value between monetary prize and recognition treatment. It suggests that in the case of TID, which can probably be extended to the corporate community in general, using non-monetary incentives as recognition and reciprocity can incentivize annotation behavior. This, in turn, can be explained by the fact that within corporate environments visibility matters more to people than immediate monetary compensation for the work done. Coherently, competing for the monetary prize does not necessarily indicate that workers will produce more for the possibility of obtaining more money; it is rather the case that the prize itself takes on a strong symbolic value.

The second obvious difference regards the value of some of the gamification features of the interface. Badges and other status indicators, such as real-time standings, become useful only in situations where building an in-game person is gratifying to subjects. Clearly, the more time individuals spend using the software, the more valuable these features become. We can therefore conclude that both time and effort spent on introducing badges, status indicators, etc., are worthwhile only when users interact with the software at some length.

At the current stage of development there are still some lingering issues and limitations which appear to be valid both for the case under scrutiny and more in general.

The quality issue has not been solved at the level of incentives. The proposed structure of scores should be related to the application scores upon which the competition is conducted.

The first issue regards the problem of assessing the quality of annotations. A few consequences of this issue emerged during the field experiment for subjects competing for the monetary prize. Under this condition a few users exploited the fact that in its current implementation the tool does not have a control impeding the use of the same annotation for the same resource. Users can submit as many same annotations as they want and each time they will be assigned points for doing so.

A second issue and current limitation of the tool, linked to the first one, is the relationship between incentives and quality of annotations. Currently, incentives are correlated only to quantity and a very rough measure of quality (we can only distinguish between tags belonging to a given knowledge base, with or without validation of their semantic contents, and free-text annotations). Ideally, calibrating scores and relative incentives to each tag generated according to a case-by-case assessment of quality would enhance the value associated to each annotation and could be used to promote virtuous behavior.

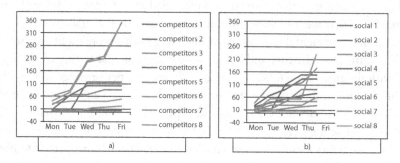

Figure 3.12: The level of contributions for a) the monetary prize and b) the social recognition groups.

CHAPTER 4

Case Study: Building a Community of Practice Around Web Service Management and Annotation

This chapter describes the process of building a community around Web service management and annotation. It further illustrates the conceptualization, implementation, and evaluation of different approaches to foster an active community around the seekda Web service portal. We utilized and extended participatory design methods and also took advantage of existing crowd sourcing services to leverage contributions to the Web portal.

4.1   AIMS AND OBJECTIVES

The emergence of the Web API paradigm allows for a completely new e-business software, which can be sold and operated in a dynamic and ad-hoc fashion. This e-business software can be enabled and at the same time make new business opportunities such as e-commerce marketplaces possible. The emerging seekda portal, in particular, collects information from Web APIs to help consumers select the right product or combination of products. The main idea is to create a marketplace, an Internet-based environment that brings together business-to-business buyers and sellers so that they can trade more efficiently online. There are three distinct types of e-marketplaces:

- Independent: in which public environments are trying to attract buyers and sellers to trade with each other.

- Consortium-based: in which sites are established on an industry-wide basis (in this case a number of key Web API providers get together).

- Private: in which e-marketplaces are established by one organization to manage their purchasing alone.

   In the C2C marketplace there are some well-known examples like eBay, Craigslist, or My-Hammer. Yet there is no respective counterpart marketplace of this size in the area of Web APIs, which connect buyers and sellers of services from disparate locations and industry fields. In this scenario, seekda's mission is to facilitate on-demand use of services over the Web. As a first step seekda is operating a search engine providing access to publicly available Web APIs. Seekda will simplify purchases across different providers and unify the use of services in bundles. Therefore, the emerging seekda portal can be a good candidate for such an independent Web API marketplace aiming to simplify purchases and transactions across different providers and to unify the usage of services regardless of their origin.

Nowadays, the current Web API market is far from being transparent. Consumers have to spend extensive resources to locate and procure services. The current market is also mostly based on atomic service offerings, when completely integrated solutions are clearly needed. Seekda's products aim at creating a more transparent and accessible Web API market. The company has developed automatic means to identify Web APIs (on the World Wide Web) and has devised algorithms to enable users to find appropriate APIs for a given task efficiently. By pre-filtering the Web content and indexing Web API specific features, seekda manages the largest set of Web APIs known and make comparison easier through a unified presentation.

As depicted in 4.1, the seekda marketplace will facilitate the trade of Web API usage in a one-stop-shopping manner—dramatically reducing procurement costs. The current market is mostly based on atomic service offerings, when completely integrated solutions are clearly needed. Seekda will address this demand by facilitating the creation of service bundles. Interoperability issues between different providers will be handled by the marketplace, which allows for a seamless switching between providers and thus reduces integration costs for the customers of seekda.

One of the still unresolved problems is the annotation system. In order to find useful Web APIs and be able to bundle them, a very rich system of annotation is needed, because the syntactic description of available Web service interfaces is insufficient for the automation of the service usage process, especially with the constantly growing number of available services. Semantic Web Services (SWS) technology uses semantics to describe Web services in a machine-readable form as a response to this issue. They are confronted with the same problems as other areas in which semantic content is created: the severe lack of semantic service descriptions. Over the past few years, a number of frameworks for describing SWS have been developed including WSMO, OWL-S, and SAWSDL. However, existing semi-automatic methods to annotate Web services associated with these frameworks are still in the beginning stages. Most of them also require training sets constructed through human labor. Manual service annotation, to create training sets for semi-automatic approaches, is currently the most feasible approach in this area. Identifying the stakeholders (users and contributors) of such a Web service search engine is therefore crucial to be able to adequately motivate them in contributing in a way which benefits all stakeholders. Having identified the different stakeholders of the search engine as well as their motivations for annotation and possible incentive, an issue arises that a search portal which is trying to use annotation mechanisms to enhance both search efficacy and user experience, cannot only focus on single user interaction and experience, but has to take the interrelations of all users' actions into account. Regarding this issue, the design and introduction of groupware systems into organizations, social networks, and communities, approaches to

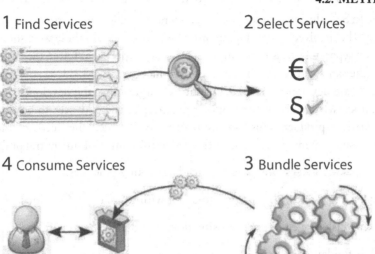

Figure 4.1: Sequence of customer interaction on the seekda marketplace portal.

socio-technical systems and Participatory Design (PD) have proven to provide appropriate methods, tools, and procedures to foster engagement and participation [12, 102]. In particular, the usability, the appropriation, and the acceptance of groupware applications have been significantly increased through PD methods and the involvement of potential users in software engineering projects. Furthermore, the success of the social Web or Web 2.0 applications (and most open source projects) is based on the establishment of online communities whose members significantly contribute to the design and the content of these social Web systems. During recent years, issues of communities and technologies, social capital, and trust-building have gained prominence in socio-informatics and community computing [57, 58]. By enhancing the seekda search portal in terms of these social aspects and by incorporating appropriate incentives for users and providers of the services to supply additional annotations, the scalability of the system, even when managing millions of Web services, is achievable because of the high quality annotations.

4.2   METHODS USED

We used the process-oriented ordering of practices that has already been described in Figure 2.3. Ideally, the process of design and development of an incentivized application should start with the analysis of the specific situation, we have focused our analysis on seekda stakeholders and their interests. Due to the fact that the portal was already being used by various individuals, we tested it directly in the field, using the following methods of analysis and followed a classical participatory design approach. The school of Participatory Design (PD) was established in the 1970s as "cooperative design" supporting a humanistic attitude to work in lieu of more traditional tayloristic and fordistic approaches. In Scandinavia, workshops for cooperative design were organized following up to workers wanting to be involved in developing processes. Originally, in these workshops, management and workers acted separately, thus maintaining a tayloristic approach. Later, the two groups participated together in the same setting on the design of products [1, 12, 13, 33, 52, 64, 67, 92, 109].

In software development, the PD approach argues that the developer does not have enough fundamental knowledge about the working practices of the end-user to be able to prepare an appropriate software product. This can result in a gap between the needs of the end user and the application development. Greenbaum and Kyng in [47] outline four central problems for IT design:

- Developers of IT systems have to consider working practices;

- They have to work with human actors, not with roles, tasks, functions, or structures;

- Working tasks have to be seen as situational actions in their context;

- Work is fundamentally social and involves extensive cooperation and communication.

Previous developmental models of software engineering like the waterfall model [14, 107] leaned on phases in the project flow and are characterized by their base of a sequential succession of activities. At the end of every phase, one or more documents, which the developers use to communicate among each other, are released, whereas communication with the end user is established at the end of the project. This procedure did not lead to optimal results, because users were not involved during the development phase. In addition, requirements rarely can be recorded in a complete manner before the development. According to Wanda Orlikowski [98], these gaps are particularly problematic because the introduction and the use of software within an organization fosters changes and the creation of new needs, which should be considered. A strict observance of a linear approach increases the risk of losing important information and ignoring the sociality of work which is crucial [47].

We used interviews, workshops, and usability tests to gather the requirements and to evaluate the prototype in each stage of the development.

4.2.1   USABILITY TEST

To identify usability shortcomings, we focused on the usability dimensions "effectiveness," "efficiency," and "satisfaction" according to DIN EN ISO 9241 Part 11 (see Figure 4.2). The user tests were conducted at the Usability Laboratory of the University of Siegen (USI). The tests were conducted by two scientific assistants of the University of Siegen.

The tests were divided into a scenario-oriented walkthrough to identify shortcomings relating to effectiveness and efficiency and a semi-structured interview to analyze user satisfaction and further user and system requirements. The users of the study had a computer science background and were used to the terminology used in software development. The method used during the scenario-oriented walkthrough was participatory observation including "thinking aloud". The tests were recorded on audio tape and in part by screen recorder. Each user exploration session was observed by

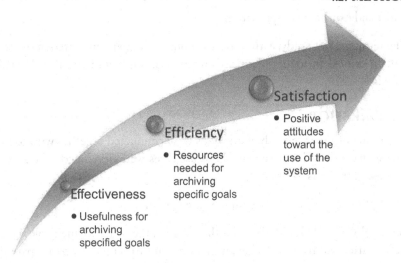

Figure 4.2: Level structure of software-ergonomic requirements.

two usability experts who asked questions and took notes. The project participant had to fulfill typical tasks and were asked to verbalize, loudly, any problems due to unknown terms or other distractions.

4.2.2 INTERVIEWS

Interviews were used to acquire knowledge about the general context of the platform and to understand the users' intentions. Further semi-structured interviews were conducted before and after the scenario-oriented walkthroughs to identify user expectations (before), to measure the users' satisfaction and thus to identify further requirements. The interview guideline is based on the SUMI (Software Usability Measurement Inventory) questionnaire and was expanded by additional questions to obtain further information about the usage context, user attitudes, and user expectations. We first conducted a semi-structured interview to find out what expectations and requirements the users have in general using the portal. Further, the participating users were asked to fulfill tasks and then to verbalize (aloud) what they were thinking, intending to do, and expecting to receive from the tool (thinking aloud method). After the observation, the users were asked questions by the usability experts, using a semi-structured interview technique. The questions related to usability and consisted of the following categories:

- Impression of the tool

- Intention of the tool

- Usability of the tool according to the dialogue principles (EN-ISO 9241-210)

- Suitability of the provided help documentations for the task

- Effort and expected compensation

The results of this study, with respect to both aspects, are presented in the tables in Chapter 4. They contain critical incidents identified in the user study and the ISO-Violation according to EN-ISO 9241 – 210.

4.2.3   WORKSHOP

In the beginning of the case study, workshops were conducted together with users, developers, and the seekda management. In these workshops use cases were developed. The aspects considered for the seekda use case are as follows:

- Description: describes the overall use case scenario.

- Actors, Roles, and Goals: identifies the actors in the scenario, their roles (i.e., what they do in the scenario) and their goals (i.e., what they want to achieve by participating in the scenario). In order to make these use cases more concrete we decided to name the characters that appear in the usage scenarios.

- Storyboard: describes a sequence of interactions between the involved actors and the seekda portal in terms of atomic steps, i.e., an activity that has to be performed. There could be more storyboards in a use case if this makes the storyboard more readable.

- Requirements: integrated into the storyboard, the requirements resulting from the use case. In a later phase of the project online workshops were conducted to include the user into the development of the portal, giving him the task of uncovering potential bugs and other areas of improvement.

4.3   CASE STUDY DESCRIPTION

As described the development of a system and later the willingness of its users to actually use it benefits remarkably from applying PD methods. In regard to this, the inclusion of all identified stakeholders is preferable. In the case of the seekda engineers, the inclusion can be easily done since they have a genuine self-interest in enhancing their own product. For all other stakeholders (service provider, general portal visitor, and visiting developer) participating is to a certain extent more complicated, due the distribution of these users all over the world. Classical PD approaches tend to gather all relevant stakeholders in workshops or to bring them together for a face-to-face discussion. Due the delocalization of the different users, we tried to move the workshop setting into a virtual environment as an Online Participatory Design (OPD) workshop. By doing this it is possible to overcome the geographical and time differences by offering an asynchronous coordination and communication platform. Still, the effort to participate in the virtual process is higher in comparison the classical approach, since the discussion is not conducted immediately and verbally, but has to rely on asynchronous, and in our case, written interaction. As a consequence, before even introducing the OPD workshop, we first conducted a requirements analysis to identify an initial set of aspects necessary to enhance the prototype. By doing this we hoped to identify and fix small issues. For the OPD workshop this would mean that participants can focus on major issues (e.g., structural problems, missing functionality, etc.), by which we tried to compensate the higher effort to communicate and discuss. The requirements came from eight interviews and a focus group with 14 participants, all representative employees of seekda and users of their Web services search engine. The results focus on the analysis of the seekda Web service search engine prototype and tend to identify design requirements, with respect to: usability requirements community requirements, and incentive requirements, as well as intrinsic motivation aspects that might drive users to contribute to the portal. Other variables that should be taken into account in order to design incentive mechanisms, such as players, are rules and social context; expected outcome and pay-offs; goals and tasks.

4.3.1   INITIAL REQUIREMENT ANALYSIS

As said before, the initial requirements analysis was conducted to reveal minor issues with the portal which nevertheless influenced the user experience. To prevent the later participants from focusing on these issues in the online discussion, we conducted a first set of user inquiries to resolve those issues ex ante to the OPD workshop. We conducted interviews with eight representative employees and experts of seekda. Each semi-structured interview was done by two interviewers, took 60 to 90 minutes, and was recorded on audio tape. These recordings were then transcribed and analyzed according to the ex-post categories. Additionally, a focus group discussion was conducted, focusing on the usage problems of the existing system and on possible design solutions to overcome these problems. The interviewees tried to explain whether and to what extent semantic annotation could actually improve the seekda Portal. Seekda interviewees' feedback was decisive for the direction of the design, depending both on their impressions and their usage along the way. The interviews and the workshop resulted in identifying several issues (>60) that needed to be addressed before the OPD workshop was conducted or at least could serve as hints to facilitate the discussion if user's activity during the OPD workshop decreased. Stating each issue would not serve the purpose of giving an overview of the process we conducted to build up a community around the seekda portal. All results can be found in [25].

4.3.2   APPLYING OPEN PARTICIPATORY DESIGN

Based on the initial requirement analysis and the experiences in the first redesign phase, it was decided to establish a Participatory Design (PD) process in the seekda use case study. This process was initiated by interviews with seekda developers and users in the first project year and two PD workshops (located at seekda venues in Innsbruck and Vienna). These interviews and workshops were aimed at analyzing user needs and at planning and designing a suitable establishment strategy for this particular Participatory Design method. The traditional PD approach has already been introduced, therefore the chosen Online Participatory Design (OPD) methodology will now be described as well and the development and implementation of the OPD dashboard as a technical infrastructure for the OPD workshops will be presented. Following the principles of PD, the dashboard was designed in a participatory way as well, improving its design significantly through the involvement of potential end users.

The implementation and most parts of the OPD concept were developed by a student from the University of Siegen. For a more detailed description on the technical aspects of the prototype, as well as a more profound description of the process and the participating user, we refer to the diploma thesis of Marc Herbrechter [56].

As already mentioned it became obvious in the first PD workshops with seekda that the end users of their Web Service Search Portal are online users, and thus we chose an online setting for our planned PD workshop(s), using Online Participatory Design (OPD) as an approach. In a PD workshop in Vienna in August 2010 experts from Siegen University and seekda developers identified requirements for the process and the technical infrastructure. Furthermore, a project plan with a time frame and milestones was defined. Coordination mechanisms were developed and responsibilities were assigned. In the following we will describe the planning and conduction of the OPD workshops and the development of an appropriate infrastructure.

Process Planning of the OPD Workshop

Our OPD workshop is conceptualized as a cyclic, iterative process, conducted either within a set timeframe or targeting specific design goals. Time constraints that are suitable for software engineering processes (even with versatile programming techniques) require that OPD workshops (including two design/redesign cycles) last several weeks (a minimum of four to six weeks) for each cycle (overall up to four months).

Stakeholders and Roles

In order to represent all of the relevant stakeholders, four roles were identified and implemented during the OPD workshop. Each role was assigned specific tasks and responsibilities:

- Project owner

- Research/Observer

- Technical committee

- User committee

Project Owners initiate and coordinate the OPD process. They usually come from the (software) company's management and possess the required competencies to make decisions regarding resources and frame the conditions for the process. They are mainly assigned to planning and coordinating tasks without intervening in the operative workshop activities of the PD process. They allot resources and are responsible for supporting the operative workshop activities. Furthermore, Project Owners send software developers to the Technical Committee. Project Owners in the seekda case study were a project manager from seekda and an expert from the University of Siegen. Together they decided to start a first OPD workshop cycle on November 1st, 2010. **Observers** were researchers. They acted as silent observers or as participant observers, following the action research approach and ethnographic methodology. The observers monitored and evaluated the OPD workshop, in order to advise workshop participants in regard to aspects of quality management. In this case, the observers were the student implementing the process as well as two researchers supervising the student. The **Technical Committee** (TC) consists of software developers and end users (ideally, in a 1:1 proportion). The size of the TC depends on the size of the project. TC members should be able to decide on important matters in the (re-)design process without interrupting the process for each decision. Technical Committee members should collaborate directly with the Project Owners to guarantee valid decisions. End users in the TC should be highly engaged in the process. In our case the TC consisted of two people, one was the lead developer of the seekda portal, the other one was responsible for the OPD prototype. The **User Committee** (UC) represents potential end users of the software prototype. In small projects all participating end users can sign up for this UC; in larger projects, the mechanisms for delegation should be foreseen. The UC reflects on experiences and needs of end users and coordinates end user activities in the OPD process. In the INSEMTIVES Seekda case study developers from the TC also subscribed to the UC. They received "first-hand" information without intervening in UC decisions. On the other hand, this involvement of TC in UC can lead to a biased interaction among users. All in all there were 15 users involved in the OPD workshop. The process was accompanied by 15 to 20 users from all four stakeholder groups. The group consisted of Web service experts, developers, and random visitors.

Process Structure
The following section describes the structure of the process to be followed during the workshop. On the macro level, three main tasks were foreseen: evaluation, discussion, and election. After the evaluation of design alternatives, changed requests were articulated and were discussed with the other participants. In the third step the features were selected and prioritized. However, on a micro level, the process was characterized by many different sub-processes, such as the creation of a mock-up for a meeting in the forum or the marking of relevant fields on a screen shot. The discussion and election of design alternatives was communicated between Technical Committee and the User Committee and decisions on specific design requirements were made consensually.

OPD Dashboard
In the following section, the specific implementation of the online Participatory Design-Dashboard (see 4.3) is presented in detail. The section is split into the system design (as well as implementation details) and the evaluation of the dashboard.

System Design
In the following, the design of the online PD dashboard is described. To design this dashboard, PD workshops with the company seekda in Vienna were conducted. During these workshops,

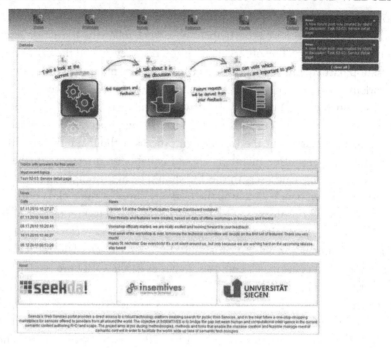

Figure 4.3: Online Participatory Design Dashboard.

requirements concerning the OPD system were specified and the application of the system was discussed critically. The participants of this first PD workshop were the project management and the developers of the seekda, as well as the project management from the University of Siegen. After the development of the first mock-up prototypes, potential end users were also involved in the process to contribute to the deployment of a first usable version of the dashboard. During the workshop, the following requirements concerning the software (to support the online PD workshop) were identified:

- The possibility of user registration

- Discussion forum for exchanging ideas and messages

- Feature list with specific functional and/or design requirements for the developer

- Stating feature requests and request voting

- A user committee and a technical committee

- Observance of the corporate design of the company seekda

To visualize the requirements of the OPD dashboard, mock-ups were created at the end of the workshop. These were similar to paper-based mock-ups and offered a rough overview of the user interface to be implemented. The main requirement in creating the user interface was simplicity. This meant that the main functions (forum and feature list) should be available from everywhere and the dashboard should follow a flat navigation structure. Through this, the user can reach every function over a maximum of two navigation levels in the final implementation. A forum was incorporated which was dedicated to the discussion of specific design solutions, features presented in the prototype section, and the feature list. All participants were able to open new discussion threads with an initial statement or a question. These discussion threads were the main place for communication and interaction between system developers/designers and end users.

Implementation of Awareness Mechanisms
Probably the most important part of the implementation was the awareness solution which informs the user in real time about events happening on the platform and about actions that were done while the user was active on the platform. With every access on a site of the dashboard, a thread is started which continuously searches for new activities to keep the user updated on other users' activities. The introduction of the notification system gave the users a "live view" on the activities of the other users. Its main aim was to enhance the group awareness. Being part of the group was supposed to function as a motivation to make sure users took part in the collaborative process of participatory design. These assumptions were based on interviews we conducted to develop motivational mechanisms for a collaborative tailoring environment. Exemplary statements are shown below.

> *"By seeing the contributions of other users I know that the whole thing is not already dead and therefore I feel motivated to contribute, too. Being aware of others' activities creates a sense of community, because everyone shares a bit of their knowledge and by contributing I can become a part of this community as well."*

> *"The team spirit leads to higher self-motivation to participate. That's what keeps the community going. If I use it, I want to give something back."* (translated by authors)

User Evaluation of OPD Dashboard
The OPD dashboard was designed with help of potential users. Four students were involved in usability tests of a first prototype. The usability tests took place in a collaborative and locally distributed setting. Users were interviewed and observed. After the user performed the test, a semi-structured interview was conducted. All observed and recorded data was analyzed with regard to the design requirements for the further development of the dashboard. To evaluate the fun factor of the OPD dashboard prototype, Attrakdiff2 was applied, a questionnaire focusing on the hedonic quality of artifacts [53]. The questionnaire differentiates between pragmatic and hedonic quality factors and offers 28 pairs of adjectives in a semantic differential. Users have to rate each pair of adjectives on a rating scale, evaluating (i) pragmatic quality, (ii) hedonic quality of stimulation, (iii) hedonic quality of identity, and (iv) attractiveness of the dashboard. Thus, a pre-test-post-test design was chosen, and the questionnaire study was repeated after the redesign of the prototype. Based on the observations, interviews, group discussions, and the Attrakdiff2-survey, specific design requirements

were derived and implemented in the prototype. The Attrakdiff2 study done before the redesign and again after the redesign of the prototype, showed significant improvements of the individually perceived pragmatic and hedonic quality of the dashboard. The redesigned prototype was rated by end users as "very attractive."

Results of the Workshop

Based on interviews with a selected group of workshop participants, an OPD workshop can be seen as a promising way to involve the user in the process of enhancing online tools. Participants from the user group, as well as participants from the technical committee, emphasized the benefits of the workshop. As already assumed before conducting the workshop, participants, who also participated in workshops for the initial requirements analysis, mentioned the lack of non-verbal communication that took place in co-located settings. Some users also criticized the performance of the OPD Dashboard.

Nevertheless, most opinions were positive. The developers of the seekda portal saw a huge decrease in their workload because of the OPD workshop: *"From my point of view it was a success. […] It helped us decrease the work of programmers, and from a planning perspective it also helped, since we had already started discussing everything in advance." (TC, Participant of the OPD-Workshop, translated by authors)*

It has to be emphasized that developers showed a higher motivation to further develop their portal because of the workshop. They felt supported while working and were provided with constructive ideas which probably could not have been implemented without the workshop. All participants highly appreciated the continuous contact via e-mail and the notifications from the dashboard itself.

> *"As soon as I visited the page, I was informed about new contributions from other users. I really liked that."* (User Group, participant of the OPD workshop, translated by authors)

> *"Really useful! There was no need to look for a post. I just clicked the notification."* (User Group, participant of the OPD workshop, translated by authors)

The participants suggested 18 new features which were derived from over 160 forum posts and ranked with over 250 votes. Six of these features already have been implemented and future development is planned. Also, several small usability and design issues have been named by the participants and will be taken into account for future versions.

4.3.3 INCREASE USER PARTICIPATION BY UTILIZING CROWDSOURCING MECHANISMS

Even though the results of the OPD workshop were promising, the development seekda portal had to address another issue. There is a need of a critical mass of adequately annotated Web services enlisted in the portal to be beneficial to the users. Trying to gain this critical mass we used a crowdsourcing service, Amazon Mechanical Turk. In the case of the seekda portal the task is not taken from the employees alone but also from the portal visitors, since in many cases they visit the portal just to find a Web service quickly. Thus, they are not motivated to comment on Web services or at least the one they were looking for, even though the effort to do so is quite small. But because of that the task is highly suitable for the micro task market [71] and a service like the Amazon MTurk. On Amazon MTurk "requesters" with a given problem pay "turkers" to solve the problem. The problem itself is split into a mostly huge number of "Human Intelligence Tasks" (HITs), which can then be processed by the turkers. Each of these HITs can be solved independently from other HITs and from other turkers. Typical tasks that can be easily dismantled in this fashion are content annotations and labeling, the ranking of various types of artifacts, completing and correcting records in a database, and many more [59, 72]. The requester groups the HITs into packages, which can be configured in several ways, and are then published on the MTurk platform. They can be configured, e.g., in terms of the number of assignments per HIT or the time it takes to complete a HIT. The profile of the turkers can be restricted in terms of geographical location, knowledge of specific natural language, and other things. If turkers agree to process a package, they are typically provided with a user interface created by the requester. It has to be emphasized that based on the packaging of HITs, the concept of the provided user interface to complete the tasks, the provided set of instructions, and explanations to process the task, the effectivity can vary significantly [32]. In our approach, we therefore first created a user interface and a set of corresponding manuals on how to fulfill the task. The users were provided with screencasts on how to use the interfaces and also with written instructions, so that they could look it up, if something was unclear while fulfilling the task. The initial user interface itself, its evaluation, and the suggested redesign, is described in the following chapter.

4.3.4   WEB SERVICE ANNOTATION WIZARD FOR MTURK

The prototype was created to guide the user in the Web service annotation process. Users were asked to distinguish between web pages, differentiating between a Web service interface and web pages that just provide information about those interfaces. For example, the official Facebook API page describes the interface of the Web service, while a blog entry about this API could, for example, just provide an example of usage and implementation. After that the users were asked to provide a description of the API, Tags, and a category. The figure below shows the initial user interface. Before the user could start annotating the Web services they were provided with help documents and a screencast showing how to use the annotation prototype.

This section now presents some of the results of the seekda's MTurk prototype usability evaluation. By showing the results of the evaluation we can provide a good overview about these aspects that are crucial for a successful crowd-sourcing approach.

Task Introduction

Users had no problems understanding the general task. Even if they were not experienced in Web developing they roughly got an idea about what a Web API is. Since the users had computer science backgrounds, they were used to the terminology of software development, but were not always sure about the special Web service terms and concepts. Therefore, we suggested explaining the aspects

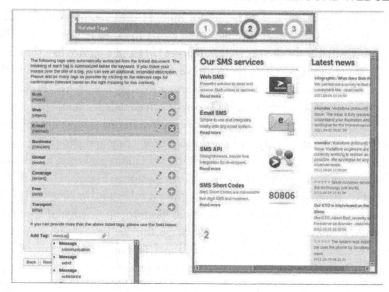

Figure 4.4: UI of the annotation prototype.

of Web service programming by comparing it with standard JAVA or C# developing. For example, users were comfortable with the term API and they knew how to use local APIs. By stating the difference between calling a local and a remote service the users were be able to grasp the concept of Web APIs.

Examples of Web Services

The examples of the Web API pages and of the false examples were perceived as useful. Still, the users remarked that instead of just providing several examples, it would have helped much more if they had received hints at aspects which revealed a page to be a false page. For example, if the page is a blog entry about the usage of the Facebook API, the description should say that this is a false example, because it is just a usage example and not the official Facebook developers' page. There should be at least one example for the most common hints for identifying true and false Web API descriptions.

Screencast

In general, the screencast was well accepted by all users. They were able to start using the tool immediately after watching the screencast. The instructions were provided only as overlays to be shown for a longer duration. Concerning this, most users remarked that they would also have preferred to have an audio description in the video, since it was difficult to read the instructions and at the same time follow the interaction with the prototype shown in the background. In the end of the screencast the result of the annotation process should be shown to the user by navigating to the seekda portal and presenting the detailed representation of a Web service listed in the portal.

Structure and Navigation
Users perceived the design well and evaluated it as suitable for the given task. Still they had remarks on specific elements of the prototype. For example, the users pointed out the importance of progress visualization. At every given time they wanted to know to what extent they had completed the given task. Other remarks concerned the typical usability issues, like allowing multiple selections in a list of options or increasing the shown part of an example web page to be able to interact more comfortably (reduce scrolling).

Setup and Results
We conducted two MTurk sessions to carry out a good design for the challenge in terms of financial rewards and needed expertise. The setup of the first phase was as described in the following:

- Initial set of 70 Web APIs for testing purpose

- No required qualification of the turkers

- Estimated time to complete one task was 15 minutes

- Payment for each task: $0.10

- Turkers were provided with a written description of the task and a screencast showing the actual procedure

  For the second phase we adjusted the set up based on the results of phase 1:

- Decrease of the time to complete one task to 10 minutes

- Increase of the reward to $0.20

- Key questions (description, tags, category) became a mandatory input to complete a task

- Strict evaluation criteria (at least one category, two tags, and a meaningful description)

- Increasing of the number of Web API to 100 example services.

    In the first phase of the challenge, we had 23 unique turkers who processed the 70 Web APIs. Spam was not a significant problem (~10-15% of the submissions). Overall 49 tasks resulted in a correct classification (Web API vs. non-Web API), whereas 21 were classified as Web API and 28 as documents not related to a Web API. In terms of completeness and quality, 15 of the submissions included descriptions, categories, as well as tags. From those complete answers, eight were of sufficient quality to be able to use them to enhance the data in the seekda portal. The first set of tasks showed that the estimated effort was too high. On average turkers completed the task in

2.5 minutes; the duration ranged greatly from 9 seconds to 13 minutes. Beside this, we recognized that only a few users added categories, tags, and descriptions. Thus, we decided to define categories, tags, and descriptions as a mandatory input for an acceptable submission. To compensate for the increased effort, we rewarded each task with $0.20.

In the second phase of the challenge we had 27 unique visitors from which only one turker had already participated in the first set of tasks. The amount of time to complete a task still ranged approximately in the same range as in the first phase. Overall only 10 tasks resulted in wrong classifications, where even five of them were web pages with high quality annotations. In the correct answers, the Web APIs were mostly annotated with two tags. The categories, which were provided by the turkers, were sufficiently accurate on an abstract level. Due to the remark that only complete submissions were to get approved descriptions, they were mostly meaningful, which in this case meant that over 80% of the provided descriptions were usable.

4.4   RESULTS AND LESSONS LEARNED

In this chapter we presented our approach on how to foster a community around a Web service search portal. After reflecting on the purpose and the current state of the portal, we described the initial analysis of potential enhancements utilizing workshops, interviews, and expert evaluations. Based on the first improved version of the portal we introduced the so-called Online Participatory Design Workshop. We described the development, first test cycles, and a short evaluation of the OPD dashboard, which is the tool we used to reproduce the co-located setting of classical PD approaches in a virtual setting. We established a stronger user involvement and were able to bring users, developers, and other stakeholders into contact, which resulted in an enhanced group consciousness. In addition, the users helped to point the further development into a promising direction. In addition to enhancing the involvement of the users, we also wanted to maximize their benefit when visiting the portal. Therefore, we used crowd-sourcing technology to create a high quality set of annotated Web services. We introduced the seekda MTurk challenge and described the process of setting up the challenge using Amazons MTurk service. To sum up this chapter we wanted to present the main achievements of the implemented strategy using our 10 guidelines to make your semantic application addictive:

1. Design your App to be Useful

   The emergence of high flexible systems was enabled by Web service technology. The services have the potential to not only address an issue once, but can be seen as best practices and therefore can be utilized in several instances of the same issues. People are always looking for the most suitable solutions for their problems, and with the seekda portal we offer the ability to find tested and reliable solutions, just by searching for similar to standard search engines.

2. Design your App for reaching your goals

   Our goal is to build a rich set of Web services and an easy way to identify the relevant Web services in this set. Since we rely on human intelligence to create high quality annotations, we brought our goals in line and the goals of our users by establishing a user-oriented design process with regard to the portal itself (OPD Workshop), as well as the process of data creation (MTurk challenge).

3. Design your App to be Usable

We spent a considerable effort on the usability of the whole platform. In several user studies, expert walkthroughs and workshops with users and developers made sure that the portal itself and tools, like the OPD dashboard or the MTurk interface, are easy to use and intuitive.

4. Design your App to be Enjoyable

During the design of the portal, we tried to stay in contact with the users to address their needs. We tried to understand why they visited the portal and for what purpose. In doing this we tried to create the best possible user experience.

5. Design your App for Visibility

By enabling the user to actively contribute to the portal, by helping during the design process or by adding own annotations via MTurk or the portal itself, we hoped to create a sense of belonging. Through their participation the user should acknowledge the portal not just as another search engine, but as the collective wisdom of the crowd. Users were asked to read, contribute, and collaborate.

6. Design your App to be Sociable

This aspect relates strongly to the previous aspect. We tried to enable all kinds of users to participate in the portal to the extent that they wanted. Users could step into the community as a "free rider" only and benefit from the contributions of others with no need to register or log in. On the other hand, more active users were free to register and, for example, join the OPD workshops.

7. Design your App to be Valuable

We tried to create value for the users by letting them decide where the development was heading. Further we offered external incentives through payment or prizes in the challenges.

8. Design your App to be Explorable

We designed the portal in a way to let the users find their own style of interaction. For more formal tasks (like the annotation process in the MTurk challenge), we provided the user with helping material to let them find out how things work and also at a pace they were comfortable with.

9. Design your App to be Flexible

From the bottom up, the portal itself relies on cutting edge technology to meet all future requirements. Such flexibility becomes even more crucial when the design process is mainly influenced and directed by the users themselves. In this case, not only does the technology have to offer capabilities to react to the tailoring wishes, but the development process and organizational structure do as well.

10. Design your App in a Participatory Way

This chapter described how we tried to include the user into the development of the portal. We were able to build up an active community to facilitate exchange and discussion between users, developers, and all relevant stakeholders of the portal.

CHAPTER 5

Case Study: Games with a Purpose for Semantic Content Creation

5.1   AIMS AND OBJECTIVES

"Games with a purpose" (GWAP) [137] is one of the most successful instances of socially inspired approaches for content creation of the past decade. The basic idea is simple, but extremely effective: tasks, which remain too difficult to be handled by computers, but humans seem to be able to accomplish them easily, are hidden behind entertaining games that target not (necessarily) experts, but primarily casual Internet users. By playing such a game with a purpose, users are (indirectly) generating data that serves the actual purpose of the game. More specifically, as the purpose of GWAPs is to advance algorithmic approaches to computer science problems, the data produced during game play can be capitalized to build knowledge corpora required for the training of the algorithms and to validate the results of such algorithms. Since the original proposal by Van Ahn in 2006, games with a purpose have been applied to tasks as diverse as image and video annotation,[1] genetics,[2] natural language processing,[3] conceptual modeling,[4] and, through our own work in the context of the OntoGame gaming framework, to ontology engineering and semantic annotation.[5]

One of the reasons GWAPs have turned out to be so popular, is the appeal these games have on people. According to recent surveys, we spend a substantial amount of time every day playing games [7, 8, 9]. Games create an environment that is in itself intrinsically motivating, offering through their very nature many motivational features such as fun, challenges, fantasy, competition and collaboration with others, and recognition, to name only a few [86]. Through games with a purpose, it is possible to leverage these features to turn the significant number of hours, willingly spent by Internet users playing, into meaningful results that lead to qualitative improvements in research and technology. The concept is particularly useful for those types of problems that so far have not been the subject of popular science, but have mainly targeted highly specialized audiences—

[1] http://www.gwap.com/
[2] http://fold.it
[3] http://galoap.codeplex.com/
[4] http://apps.facebook.com/conceptgame/
[5] http://ontogame.sti2.at/

due to the expert-driven and knowledge-intensive nature of the underlying tasks to be performed, the novelty of the questions raised or both.

There exists a wide variety of different game types and genres, each with diverse target groups, interfaces, complexity, story lines, and playing mechanisms. Equally diverse are the definitions of games one can find in the game-engineering literature. In [3] one learns that *a game is a way to look at something, anything*. Thus, determining whether something is a game or not is highly situational. This is very much in line with the emerging trend of "gamification," which denotes the application of game elements to non-game contexts, in order to motivate users to engage with an application. Zimmerman and Salen [146] offer a more constructivist view on this concept; their definition lists a number of core features of games: *A game is a system in which players engage in an artificial conflict, defined by rules, that results in a quantifiable outcome*.

This definition emphasizes several key aspects of games. The drive to become better at a certain task and to compete with others are essential motivators of game players; from a game design point of view it is essential for the game designer to define clear metrics to measure players' performance. Later on in this chapter we will discuss some of the challenges raised by the application of GWAPs to semantic data management that are related to these aspects: the definition of an optimal scoring and rewards mechanism as well as the composition of levels in games. Independently from the difficulties encountered in setting borders around the concept, authors agreed that entertainment is a key ingredient of games, in particular when it comes to casual games or games with a purpose. In this chapter we will look into these games, whose purpose is related to semantic data management activities in terms of their main components and challenges associated with the development of new games targeting specific activities. Hiding Semantic Web-related tasks behind games seems promising considering the sheer amount of hours that are spent playing games online by gamers every day. GWAPs are designed in such a way that cooperation is the dominant strategy, in other words only consensual solutions are awarded with points. This offers not only a simple form of input validation and spam prevention, but fits well with one of the core ideas behind semantic data management, those of shared conceptualizations and interoperability. This feature applies mostly to multi-player games; single-player games may contain similar effects by having the same challenge tackled by multiple players in individual game sessions. The sheer number of Internet users who might become interested in a GWAP is a resource with unique potential. If one manages to reach, on average, 50 individuals at every moment of the day and run a game for six months, the players will contribute 216,000 hours of intellectual work to the "purpose" of the game. If we assume an average wage of 10 USD is spent on conceptual modeling—which is likely much less than actual wages—a GWAP would produce semantic content worth more then 2,000,000 USD on the traditional labor market. One must mention at this point that games are not the only efficient option to solve human-oriented semantic data management problems. Virtual labor markets leveraging the crowd may offer a viable alternative, especially for those scenarios where users cannot be expected to perceive the tasks as entertaining; a typical example is when the knowledge domain that the task tackles is not very appealing to a great number of users. In [125] we compared Amazon Mechanical Turk and GWAPs with respect to the costs related to the development and operation of the approach and the accuracy of the results produced. The aim of the two experiments was to build and extend a given ontology. The findings of the comparative analysis revealed that Mechanical Turk can lead to considerable improvements along the first axis at a comparable quality of the conceptual modeling choices.

The amount of data that can be potentially generated through games is impressive. In one of the games of the OntoGame series, we observed an average of about four conceptual choices per game in a round lasting about two minutes – this means that two players produced about two conceptual modeling decisions per minute or one per player per minute. In this way we could roughly gather more than 12,000,000 conceptual modeling choices in half a year—the real number could be very much lower, due to the fact that every conceptual modeling question is subject to more than one game round, and the accuracy of the answers is realistically less than 100%.

As discussed in Chapter 1, the area of semantic data management has many characteristics that make it an ideal source of "purposes" of games. Nevertheless, despite early success stories and promising prospects [114], the application of such crowdsourcing approaches comes with a trade-off. Putting aside the (important) question of user motivation, addressing a priorly unknown user base that reaches far beyond the scope for which the dedicated tools in semantic data management were commonly created constrains the complexity of the tasks that can be feasibly subject of a game. A great majority of the tasks that have been an active subject of research and development over the past decade in semantic data management are too complex and unstructured to be turned into an entertaining game experience. The same goes for the domains for which knowledge can be reliably collected from a mass audience. This means that the ontologies and semantic annotations that result from GWAPs are likely to be lightweight (with respect to expressivity) and that they predominantly cover mainstream topics that a large part of Internet users can relate to. Going a step further, not every simple task in a largely accessible domain can be turned into a game. For this to happen, it is necessary to decompose the work into smaller chunks that can be carried out (to a certain degree) independently and to combine the partial results. In these cases, a more refined human-computation approach, which employs games and possibly also other types of crowdsourcing in order to solve individual aspects of the overall problem, is necessary. Evaluation and quality assurance methods need to be devised in order to automatically identify correct answers and to prevent cheating [65, 87, 110, 123, 132, 133]. In the particular case of games with a purpose—as opposed to similar crowdsourcing approaches such as microtask platforms—an important challenge is the design of the actual games. They have to fulfill the highest usability expectations, be enjoyable and engaging to ensure players retention and, as a side effect, serve the purpose for which they were actually created.

To conclude this introductory section, the concept of games with a purpose has rapidly achieved great resonance in many research communities confronted with the problem of attracting user involvement for the resolution of various tasks of a technical nature, which for a long time have been the realm of expert audiences. This chapter will discuss the main building blocks of GWAPs for semantic data management, provide an introduction to a selection of such games, and give guidelines for building new games. As we are talking about an emerging research field, many issues still require further investigation in order to gain a thorough understanding of the new challenges that arise when applying a game-based approach to semantic data management and to optimize the outcomes of existing games and their exploitation.

5.2   METHODS USED

According to the framework introduced in Chapter 2, games with a purpose for producing and managing semantic data can be characterized as follows:

- The goal, from the perspective of the designer of the game, is resolving a challenging computational problem. This problem is related to the activities introduced in Chapter 1. Although it is not always clearly stated in the documentation of the game, at least in the selection of the semantic games available so far, the great majority use the resulting data for research purposes. With gamification gaining popularity in commercial settings—for instance, for marketing purposes—this state of affairs could change in the future. The participation level in defining the goal of the game is also very low; typically players are neither involved in the definition of the game and the game rules nor do they have the chance to influence the way in which the data resulting from the game play is used. A shift from the research environment to companies may change the actual motivation of casual gamers and influence their willingness to participate in activities they perceive as incompatible with their own views. Alternatively they might expect additional incentives to play the games, no matter how fun and entertaining these are. For the time being, however, the designer assumes that users will engage with the game primarily due to the game experience, and only peripherally because they are inspired by the actual purpose of the GWAP.

- The tasks addressed by GWAPs can be very diverse. Typically, they are very narrowly scoped, which is a direct consequence of the way the games are designed—casual, requiring just the right amount of intellectual challenge, and fast paced. The tasks tend to be repetitive and highly specific; the same kinds of questions are asked in different instantiations in each game round. This creates a certain learning effect, but in the same time can spoil the user experience. In addition, when the assignment of tasks per player (or pair of players) does not take into account the skill level of the players and their previous experience with the game, the game designer not only wastes valuable time and human-computation resources, but even risks boring the users. This is also the case when the game is a means to validate the outputs of automatic algorithms; if the level of accuracy of these outputs is too low, the corresponding challenges are not meaningful. As games are expected to be accessible and appealing to a large audience, the required skills are not particularly high, though one way to achieve player retention is to ensure that they have the possibility to improve their skills in time. At the same time, it is essential that the actual task solved through the game challenges is not too difficult and is phrased and constructed in a way that is adequate to the game audience, which has no knowledge about semantic data management.

- The social structure in this scenario is hierarchy-neutral, though typical game elements such as points, badges, and leader boards differentiate users. However, the way these rewards are achieved is totally dependent on the actual performance of the players in the game and is not pre-defined in advance.

- As outlined in the discussion regarding the goal of the exercise, right now, most games mainly serve research purposes. Though the data are not directly accessible, in many cases they can be obtained from the game developers and reused in research projects and other initiatives. As such, the nature of the good produced is rather non-rival and non-exclusive.

From this analysis, it becomes clear that the choice of the task to be addressed by a semantic game is a key factor for the success of the overall exercise. We will further elaborate on this issue later on in this chapter, in Section 5.4.

To design a game with a purpose and to evaluate its success, one could apply the methodology introduced in Chapter 2, whereas the most important phase is related to the evaluation of different types of game mechanics and their effects on the types, the amount, and the quality of data produced and of the user experience. As GWAPs are still an emerging field in the Semantic Web community, most of the existing GWAP instances are developed merely as feasibility studies showing that a specific task, for instance, the learning of concept hierarchies, could be approached through a game. Little is known about the reasons behind the choice for a given set of game elements, and the effects of these choices still remain largely unexplored. The methodology proposed in the following chapter explains what type of methods from social and economical sciences could be used by game developers to decide how to design their incentive mechanisms, in particular, what types of contributions to reward, how high these rewards should be, and in which scenarios to rely on motivations like competition, reputation, and community. As for the games presented in this chapter, such insights are not yet available and we will not be able to elaborate on the interplay of different game elements in the specific examples. A comprehensive example of the application of the methodology was, nevertheless, introduced in Chapter 3.

5.3   CASE STUDY DESCRIPTION

In this section we will first give an overview of the main questions a game developer or semantic technologist will need to deal with in order to design a GWAP for semantic data management. Then we will present a selection of GWAPs from the literature, addressing a range of semantic data management tasks.

5.3.1   CORE COMPONENTS OF GWAPS

The first and foremost issue to be answered by a game is the actual *purpose* and, related to this, the *technical task* the game will tackle. This task is described as a series of interconnected *questions* (or *challenges*), which the players will need to answer during the game. The questions depend on the input, which is taken from the knowledge corpus the game processes and may be closed or open scale. The problem solved through the game needs to be highly decomposable into tasks which can be approached in a decentralized fashion by answering different questions. The questions themselves can have a varying level of difficulty, but in general it is assumed that an average player will be able to answer them in a matter of a couple of minutes.

As an example of a task and its decomposition in questions, consider the task of defining relationships between classes and entities in an ontology. It is assumed that the corresponding class hierarchy and instance base are already known, they form the so-called "knowledge corpus" of the game. The types of that could potentially hold between classes and entities are pre-defined as well— though one could imagine scenarios where the user defines them as well. One way to define the individual questions is to ask the user to validate triples of classes or entities and properties. Each question will pick a triple candidate and offer a yes/no choice to the user to answer. Alternatively one could ask the user to pick the most appropriate class or entity in the subject or object position of the triple, or to pick the most relevant relationship. In all these cases, the choices a user has are known in advance. For free-text answers, game developers may opt for input agreement games [77]. There are also situations where the task cannot be directly decomposed into a number of similar questions, as in the previous example. Take for instance the annotation of a collection of landscape images according to several dimensions: location, types of landscape, content of the image, etc. For such tasks, the game would have to issue several questions for each media object: where was the picture taken, what does it show, and so on.

The *input* used to generate the questions needs to be verbalized and translated into simple, unambiguous questions. When the input is generated automatically, for instance because it contains results computed by algorithms that need to be validated by the players, it is essential that the quality of these results is not too low—otherwise the game experience will be less entertaining, as many of the questions will probably not make any sense at all.

One important aspect to be considered is how to *assign questions to players*. The basic approach is to do this randomly, though optimizations are possible when information about the performance of players in previous game rounds is available. In such situations, one could avoid asking the same questions to the players multiple times, or would even adjust the difficulty level of the questions to the players' skills and experience. It is not always easy to determine the difficulty level of a question. This issue is highly task and domain specific and requires knowledge about the problem space that might not be available in the most general case. As an example of how such assignments could work, in [133] the authors introduce several word sensing games and propose heuristics for identifying trickier questions. Seneveratne and Izquierdo [110] introduces a Bayes model to decide what kind of annotation question to ask a player, depending on the previous performance of this player. They differentiate between challenges that have already been solved by other players, where new input will confirm annotation candidates and the player reliability, and new challenges, for which the correct answer is unknown.

In addition, in a multi-player game, the game developer might want to customize the ways *players are matched* to play against each other. Assuming that they can be distinguished through their

IP addresses and certain dimensions of their user accounts, it is important to ensure that the pairs of players are as diverse as possible and that pre-recorded game rounds can be played when not enough players are logged into the game at the same point in time. There are also multi-player models of games where the players are given different roles, as, for instance, in Peekaboom [138]. The aim of this game is to locate objects within images. Players are paired randomly and cannot communicate with each other, except through the game. One player is "Peek," the other one is "Boom." Boom sees an image and a tag describing it and reveals parts of the images to Peek, who has to guess the actual tag. In this way one obtains more concrete information that refers specifically to a part of an image—as opposed to the image as a whole.

The *output* of a game with a purpose is an aggregated, manually cleansed form of the players' answers. In a first step, the game developer needs to be able to *match the contributions* of the players; this is a not an issue for those challenges in which the set of allowed solutions are known, but it is less trivial in open-scale questions, where one has to deal not only with different spelling and spelling errors, but also synonyms and the like. As mentioned earlier, input-agreement game models may provide a workaround for this problem—as shown, for instance, in the TagATune game [77]— though a closer look at the output of the two games reveals that they actually solve slightly different tasks. In a selection-agreement mode, one aims to collect descriptions of music files that both players identify with. In an input-agreement game, the tags acquired are not explicitly confirmed by each player, but indirectly allow the players to determine whether the music track they are listening to is the same. Selection agreement might nevertheless work in open-scale scenarios. In fact, guessing what's on someone else's mind when seeing an image turns out to be one of the most appealing features of games like Verbosity [132]. The crucial issue here is that the number of (free-text) choices is relatively constrained, though not explicitly limited by the game developer, and that semantically equivalent player inputs can be computed at run time.

As a second step, the game developer needs ways to be able to *identify correct answers* with as little effort as possible, ideally automatically. Since individual challenges solved throughout the game are actually relatively easy to solve, the effort (expressed in time or duration) required to manually validate the inputs of the players is likely to be comparable with the effort the developer would have invested handling the task herself—thus, the need for automatic quality assurance. This includes means to identify cheaters—cooperation game models offer an effective mechanism to tackle this problem. Quality assurance techniques based on statistics complement the picture [59]. In addition, one typically incorporates redundancy in the game and defines a threshold by which, if an answer has been confirmed by different players a certain number of times, the game developer can then assume it is correct with a high reliability.

Once accurate answers are identified, the game developer needs to translate them into a *semantic format*, using existing ontologies and Semantic Web representation languages.

Games with a purpose can incorporate a wide range of elements meant to influence player's behavior and encourage engagement. The most basic game elements are *points* and *leader boards*. This assumes that the game developer is able to identify correct answers at run time, though there are several examples where correct answers are computed offline and the user is notified of the rewards after playing the actual game [124]. Scoring can be further refined if the developer has knowledge of the difficulty level of the questions or when she wants to influence player's behavior in a specific way. A typical example is, in a context of an annotation game, tags that have been already been validated by being confirmed by different players in a sufficiently high number of game rounds. The aim of the game is the collection of a large number of different tags and as such one could award bonus points for tags that have not yet been submitted by other players, but are nevertheless consensual, in order for the game designer to have a way to check that they are not just new, but also are most likely relevant [132]. Points can be associated with badges and other gadgets, attesting to the players' skills and capabilities. Achievements can also be shared within a social network, the most common realization being the integration of the game into platforms such as Facebook. These game elements leverage the categories of motivators discussed in Chapters 1 and 2.

In the following we will illustrate how existing games addressing the tasks discussed in Chapter 1 implement these core elements. The chapter offers only a very limited sample of the games with a purpose available in the Semantic Web field. A more comprehensive survey is available in [124] and at `http://www.semanticgames.org`.

5.3.2   SPOTTHELINK

Task

The goal of the game is to relate the concepts of two ontologies to each other. In our example, the objective of the game is to match two generic ontologies, DBpedia[6] and PROTON.[7] A comprehensive description of the game, including user studies and evaluation is available in [123]. DBpedia is a community-driven effort aiming to create a machine-understandable knowledge base containing a selection of some of the most important entities described in Wikipedia. In the version used in the game, the knowledge base is structured with the help of an ontology containing 272 classes and 1,300 properties, according to which millions of Wikipedia articles are classified. PROTON was designed as a so-called upper-level ontology that captures generic concepts and their features (see Figure 5.1). It is used as a knowledge backbone for information extraction and semantic annotation purposes [121]. It has a similar size as the DBpedia ontology. The game is available both as a standalone[8] and as a Facebook application.[9]

Input and Output

When a player starts a new round, she is assigned a random partner who is logged in to the game at the same time. The team has to solve challenges together, i.e., players will only get points when they give consensual replies in each challenge. Each game round consists of two challenges. First, the players are presented with a random concept along with a short description and an image (if available)

---

[6]`http://dbpedia.org/`
[7]`http://proton.semanticweb.org/`
[8]`http://ontogame.sti2.at/games/`
[9]`http://apps.facebook.com/ontogame`

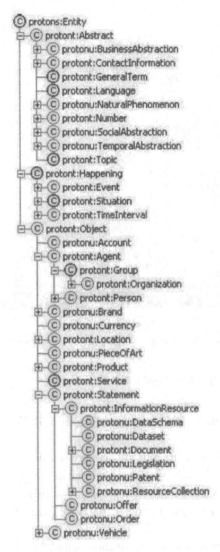

Figure 5.1: Excerpt of the PROTON ontology.

from DBpedia. In the first step, they have to choose and agree on a concept from PROTON that is related to the DBpedia concept (Figure 5.2). If the player intends to answer the challenge, she hits the "OK" button, otherwise she continues with the next DBpedia concept after hitting the "Skip" button. Only if both players agree on a decision, that is, both answer or both decide to skip, the game is continued. Otherwise, players have to answer again. In the second step, they have to agree

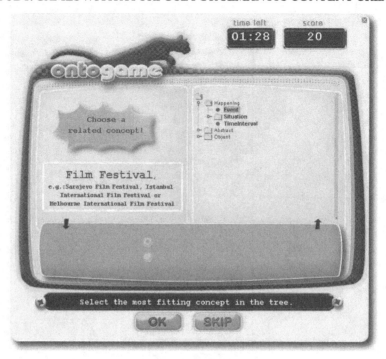

Figure 5.2: SpotTheLink: Agree on a concept.

on the type of relationship among these concepts (Figure 5.3). The process of negotiating an answer remains the same.

The outcomes of the game are alignments between the two ontologies encoded as SKOS.[10] The game output includes selection of concepts as well as the type of relationship defining the level of similarity between the concepts: narrow matches, for a pair of concepts in which one concept is more specific than the other one, and exact matches for pairs of concepts that are perceived to be the same or equivalent.

Game input is converted into ontology alignments according to the following procedure:

- First the program checks whether the concept has been played at least six times by different players.

- If that is the case, it calculates the percentage for each unique answer combination, i.e., other concepts and relationships. Answer combinations that include concepts from a deeper hierarchy level are weighted higher because they implicitly state that concepts that are located upward on the path to the root of the hierarchy tree have been agreed upon as well.

---

[10]http://www.w3.org/2004/02/skos/

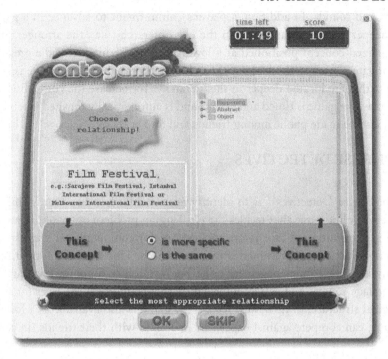

Figure 5.3: SpotTheLink: Choosing the right relationship.

- If the percentage of an answer combination exceeds 50 percent the concept match will be marked for export.

With the help of the OntoGame platform, which we introduce in more detail in the next section, each of these steps can be adjusted according to the specific behavior the game designer intends to encourage.

Incentives

When dealing with unknown input data and players, the game designer has very limited options to verify whether the players' answers are correct or not. To encourage the intended user behavior—that is, the provision of correct answers to the two types of challenges the game raises, SpotTheLink applies a customized scoring scheme. Points in games are a simple tool to motivate people to keep playing. It allows them to easily compare their performance with that of others and nurtures competition. Not doing anything or skipping game rounds does not result in receiving points, thus any player who wants to gain points is incentivized to interact with the game in the intended way. Agreements between alignment choices are rewarded with points. Due to the fact that players are randomly paired and (in theory) do not have any way to communicate outside of the game, there are low chances for developing a cheating strategy that yields positive scoring results across multiple game steps and rounds. In addition, a player's commitment to advance in a game is rewarded as well; the deeper players advance within the class hierarchy, i.e., the stronger the consensus on a more specialized concept positioned at a lower level in the hierarchy, the greater their win. This should motivate players to think about their choices thoroughly, as the score is significantly higher for concepts that are located deeper in the hierarchy. To increase the competition and encourage a fast paced play, the game is timed and each round is automatically finished when the time is up. The results can also be made public among friends and on Facebook.

5.3.3   PHRASE DETECTIVES

Task

The aim of Phrase Detectives[11] is to identify relationships between words and phrases [20]. It is a multi-player online game that tackles the challenging problem of anaphora resolution. Anaphora resolution systems are useful for many applications for computational linguistics, for instance, in information extraction, text summarization, and search. This type of information could also be of use in many semantic data management tasks. For instance, in ontology learning, one might process natural language sources and extract information that could form the basis for the development of an ontological structure. In its newest release the game is also available as a Facebook application, where people can compete against experts or in a team with their friends (in selection-agreement mode).

Input and Output

Phrase Detectives uses texts from Wikipedia, as well as fairy tales and other types of texts. From this knowledge corpus, users are assigned text fragments that they have to annotate according to specific rules and instructions. Players have to complete different tasks, such as deciding whether or not a certain phrase has already appeared before in the game, determining whether a phrase is referring to something else, or whether this phrase is a property of another phrase (Figure 5.4). Contributions are rewarded with points and upgrades to the so-called "graduate" level. Leader and top score boards show the achievements of the most successful players. Input validation is realized via comparisons with existing gold-standard texts and with answers provided by other users on the same text. The outcome of the game is a text corpus annotated with anaphoric references.

Incentives

There are three main incentives for playing the Facebook version of PhraseDetectives. The first one is to compete with other players and obtain a top position in high-score lists. The second is to take part in improving the future generation of language technology, which is explicitly mentioned in the game. The main incentive probably would be the fact that users can also win monetary prizes, depending on how well they perform in the game.

---

[11]http://anawiki.essex.ac.uk/phrasedetectives/

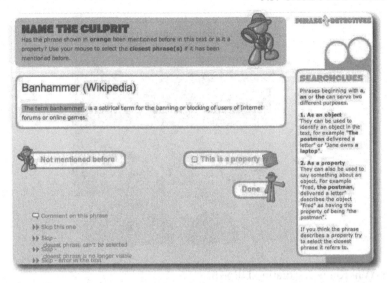

Figure 5.4: Phrase Detectives: Identifying relationships between words and phrases for anaphora reso-lution.

5.3.4   WHOKNOWS?

Task

In WhoKnows? (Figure 5.5) players evaluate the accuracy of DBpedia knowledge by answering questions automatically extracted from DBpedia triples. Questions which are marked as meaningless by a certain number of users point to data quality problems in DBpedia and could subsequently be used for ranking and curation purposes. The game is available as a Facebook application.[12]

Input and Output

WhoKnows? uses questions of the *"object is-property-of subject"* which are generated automatically from DBpedia triples. Each challenge is conceived as a multi-choice question consisting of correct and incorrect answers, which are created automatically from DBpedia triples that hold the same property but a different object. The player is provided with questions that he should answer correctly and as quickly as possible. Each question has a fixed set of answers, including correct and incorrect choices. The result is a set of validated DBpedia assertions.

Incentives

The main incentive for playing WhoKnows? is to reach the top of a high-score list, so competing with other players is a motivational factor in order to make it to the top. Another is to reach the next higher level, which also increases the difficulty of the questions to be answered. A higher level

---

[12]https://apps.facebook.com/whoknows_/

Figure 5.5: WhoKnows?: Curating DBpedia.

can be reached by giving correct answers, thereby making the game more challenging. You can go down a level by giving the wrong answers. To make the game more exciting, the user has a limited number of lives, which means that the game ends when there is no life left.

5.3.5 MATCHIN

Task

Matchin (see Figure 5.6) is a game for image annotation. The main goal is to discover pictures on a given topic that a high share of users seem to prefer. Image annotation is a classical scenario for GWAPs, given the limitations of automatic techniques and the obvious superiority of human computation when dealing with understanding images. Many examples of image annotation games exist [124], each of them using a slightly different game plot. In the ESP Game [132] players consensually agree on tags relevant for the shown image, possibly excluding tags on a taboo list. This is an example of a selection-agreement model which leads to high quality, diverse annotations. FlipIt[13] is built as a version of Pairs, where players click images to turn them upside down and then recall the ones that are similar.

Input and Output

Each game round starts with two random images from the Web. Both players have to choose the one they like better in real time. If they select the same they earn points and another pair of images is shown. Agreeing on the same images more than once in a row results in a higher reward. The longer the sequel lasts, the more points the players gain. If a sequel of consensually selected images is

---
[13]http://www.gwap.com

Figure 5.6: Matchin: selecting the best image on a given topic.

interrupted by a mismatch, the reward is reset to its initial value again. Matchin produces very high level semantics for images, namely the common feeling of many people toward the pictures shown. Input for this game are photographs. The game validates the generated content by consensus on the one hand and by the number of answers on the other hand.

Incentives

The game elements implemented in Matchin are similar to the ones that can be found in any of the games of the GWAP suite: intellectual challenge and social interaction. This is a result of the two-player model and the rules of gameplay that enforce a consensus, as well as competition and reputation promoted through points, prizes, and leader boards.

5.3.6 UNIVERSE GAME

Task

The aim of the Universe Game is to annotate images with DBpedia concepts.[14] The game is conceived as traveling in a space ship through the universe. During this journey the player encounters different types of challenges. Some of these challenges purely serve a game plot motive, others are directly related to the actual purpose of the game. As such, the Universe Game is at the intersection between casual online games as we know them and GWAPs, whereas the second has only a minor impact on the overall game story and play. The advantage of this approach, compared to pure GWAPs, is the appeal of the game to casual Internet players. Many GWAPs tend to focus on the actual technical task and spend little time on designing an exciting game story, thus hampering with the game experience. Most GWAPs are built as quizzes in which players are repeatedly asked to answer questions, some of them meaningless, as they are generated from the output of some automatic technique and their success has been limited. Experimenting with scenarios in which the actual game plot and the experience play a stronger role—at the expense of solving only a small bit of the actual purpose of the game—might provide an alternative way.

Input and Output

The Universe Game is a single-player space strategy game. The player chooses a battleship with pre-defined characteristics, deploys his supply bases, collects resources such as fuel, and scans for hidden enemies to destroy them. In his quest, he is now and again confronted with small challenges, in which images need to be classified according to a pre-defined set of categories, such as asteroids, galaxies, or stars. With each quest and each challenge he earns points, and a leader board lists the players with the best performance. There is also a bonus system encouraging more detailed annotations. The annotation component is based on a library of several hundred astronomical images collected for this specific purpose. The annotation process is two-tiered, allowing for both free annotations and a right or wrong answer. The data resulting from the game is available as a Linked Data data set in the CKAN repository, which is commonly used for the publication of such data sets.[15]

Incentives

The Universe Game implements a wide range of game elements, including a sophisticated points scheme rewarding deep-level annotations, different levels of difficulty, and upgrades to encourage player retention [45]. As the skills of the players increase, they can upgrade their battleship and gain access to more challenging levels of the game in terms of the number of depots, the configuration, the difficulty of the space maps, and number of objects to be found. The points earned in the previous round can be selectively deployed on any given spaceship to enhance performance in one of three categories—probes, bombs, and radar range. Thus, the game play is to a large extent open-ended and can be extended.

[14]http://www.universe-game.net/
[15]http://thedatahub.org/dataset/insemtives

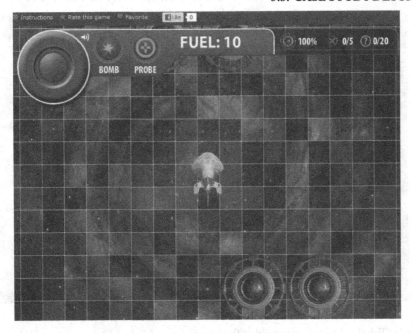

Figure 5.7: Universe Game: Image annotation as part of a space battle.

5.3.7 TUBELINK

Task

The purpose of TubeLink[16] is to create video annotations using a predefined collection of tags. In this particular case the player watches a video in a crystal ball, which contains a collection of tags to be chosen for annotation (Figure 5.8). The player drags-and-drops the tags onto the video and when a sufficient number of tags is reached, the video "explodes."

Input and Output

The tags used in TubeLink refer to concepts from the Linked Data Cloud, a selection of potentially relevant tags is computed automatically, thus illustrating how human and algorithmic computation can be interwoven to solve the given task. The game works in a single-player mode and results are compared to those of the other players to determine the scores. The videos are chosen from YouTube and the categories available there are used to identify data sets that can serve as potential annotation vocabularies. For each annotation the game records the number of players confirming it, the time of the video for which the tag is likely to apply and the corresponding Semantic Web concept. The results are available as RDF.

---

[16]http://ontogame.sti2.at/games/

Figure 5.8: TubeLink: annotating video fragments using Linked Data concepts.

Incentives

TubeLink implements a novel game plot for video annotation along a number of common game elements such as scores and leader boards to create a pleasant game experience.

5.4   BUILDING NEW GAMES

In this section we will introduce the Generic Gaming Toolkit, an extensible framework that provides the main building blocks for the design and implementation of semantic games. Furthermore, we will discuss a series of general guidelines derived from relevant literature on games and game mechanics, as well as from our own experiences in the context of OntoGame and the INSEMTIVES project.[17]

5.4.1   THE ONTOGAME GENERIC GAMING TOOLKIT

The Generic Gaming Toolkit consists of a programming API that covers the most common functional components, which are required to turn a specific technical task into a game with a purpose and a technology platform to run such games on [116, 122]. Both the source code and the associated code documentation are publicly available at SourceForge.[18]

The Generic Gaming Toolkit makes a conceptual difference between challenges, inputs, resources, and play records. A *challenge* abstracts from the actual task players carry out while playing the

---

[17]http://www.insemtives.org
[18]http://insemtives.svn.sourceforge.net/viewvc/insemtives/

game. Each challenge can be split into several sub-challenges, which are structured hierarchically. In SpotTheLink, for instance, the main task is to align concepts from two different ontologies. This task is further divided into two atomic tasks: deciding whether two concepts are related and finding the appropriate relationship capturing this correspondence. A *resource* abstracts from the topic a challenge is about, for example, a YouTube video, an ontology concept, or a Wikipedia article. A resource may also refer to the answer provided by a player, which in the case of SpotTheLink are classes in the DBpedia and PROTON ontologies, as well as SKOS relationships. An *input* represents the answer of a player to a certain challenge. It is always related to a resource, deals with a specific challenge, and may contain a resource as answer. A *play record* behaves like a container for player inputs on a specific resource. It is used for all subsequent background computations through which player inputs are translated into semantic content, including the evaluation of the inputs and the actual encoding using Semantic Web standards.

Partner Matching

A core feature of the Generic Gaming Toolkit is the matching of the players in multi-player games. When a users logs on to a game, the system selects a partner according to criteria such as the IP-addresses, location, age, and gender—all of which may influence the quality of the inputs collected. The selection of players within the game ensures that players remain anonymous while consensually resolving challenges, thus increasing the probability that they behave as intended by the game designer. Other games from the OntoGame family also use recorded game rounds, for which answers are already available from previous players, in order to realize a single-player mode in addition to the two-player mode deployed in, for instance, SpotTheLink.

Players' Reliability

Reliability is a rough indicator of the trustworthiness of the answers provided by a player. Having such an indicator is a very important aspect of the process of semantic content creation in the Generic Gaming Toolkit, since the game mechanics are always about questions where the answer is actually unknown to the game designer. It has to be assumed that there are at least some players that want to trick the system and behave in an irrational and unfavorable way. As the very purpose of the games is to derive useful structured knowledge from game inputs, each game needs to implement functionalities to verify whether an input comes from a trustworthy player or a cheater. Permitting cheating would decrease the gaming fun as well as the quality of the generated data.

Consensus Finding

One of the most important features of the Generic Gaming Toolkit is the abstract projection of the process to agree on an answer from the options given for each question. Finding consensus can be achieved in many different ways, such as by a simple majority. However, there are also more sophisticated ways of identifying a consensual set of answers, e.g., by weighing each answer by the player's reliability, considering only answers that have been answered in a similar way by all players of a certain game round. Additional restrictions, such as a minimum number of different players, or a minimal number of answers on a challenge, can be imposed on the process of consensus finding.

Ranking

Ranking is used to give an indication of the importance of a play record. The importance defines which record is most suited to be played next. Ranking metrics can be implemented in various ways, depending on the needs of the game, i.e., one could rank the records that are closest to produce the highest output or records that have not yet been played very often. The Generic Gaming Toolkit uses two different strategies: one that does not rank at all, which is required when ranking is not supposed to be considered during the record selecting procedure, and a second one considering the number of required answers, the number of different players, and the number of questions that have been answered.

Matching

The last conceptual feature of the Generic Gaming Toolkit is definition of identical answers. Note that this is different from assessing consensus, as matching answers are used to compute equality between answers. An example of doing so is comparing the answers of two players and evaluating their similarity. An implemented matching algorithm checks whether a player's answer equals the majority of all previous collected answers on this resource.

5.4.2   DESIGN PRINCIPLES AND OPEN ISSUES

In this section, we will summarize some of the most important lessons learned over the last three years of developing the OntoGame framework. The ultimate aim of this research is to provide comprehensive decision support in matters related to the execution of human-aided semantic content authoring tasks. This can be achieved by identifying the tasks that can be effectively addressed through games, crowdsourcing platforms such as Amazon Mechanical Turk, and social platforms.

Task Selection

The identification of the semantic-content-authoring tasks that are suitable for the casual-game paradigm is of topmost importance for the overall success of the approach. Hiding tasks behind games is not straightforward, and cannot work for every aspect of the overall process, no matter how highly human-driven it might be. Candidate tasks cannot be too difficult or too easy, but they have to be divisible or combinable, so that they can be broken down into smaller chunks that can be solved independently by a potentially large group of contributors [117]. They have to be suitable for a broad audience of players and, in the context of games, be mappable to a series of consensual decision-making challenges. Our experiences show that the resulting workflows do not have to be constrained in just their structure—basically the sequences of atomic tasks which are approached and solved consensually by players—but also in size. As a rule of thumb, a sequence of more than three to four interrelated tasks is likely to lead to challenges that are too complicated for players to learn and keep track of at the fast pace the game is expected to be played in. This makes the game less appealing for players, thus diminishing their willingness to play and reducing the amount of data produced. Each atomic task results in a question—to which the correct answer is not known in advance to the game designer—which needs to be answered by the pair of players competing against each other in a pre-defined period of time. This includes open-scale questions, whose answers are typed in by users, and closed-scale ones, for which the system provides a set of possible answers from which the players have to choose one viable option. In the first scenario it is necessary to make use of specific matching algorithms to cope with potential variations in the form of the answers, different spellings, for instance, in order to ensure that consensus finding is possible in most cases. Nevertheless, if the set of potentially correct answers is too broad for a consensus to be likely, the task is probably less appropriate for a game-based approach and additional knowledge has to be taken into account to reduce the amount of possible solutions. An interesting research question in this context would be the extent to which it is possible to combine a games-based approach, which can obviously solve only very specific types of semantic-content-authoring tasks, with other human-computation paradigms and incentive mechanisms. For example, one could imagine refining the results of casual games in Amazon's Mechanical Turk or different combinations of games similarly to the GWAP framework.

In many areas of semantic data management, a game-based approach is realizable as a means to validate the results of automatic algorithms carrying out a specific low-level tasks, such as extracting entities from a text corpus as a foundation for building ontologies or creating RDF data sets. Each of these aspects, however, has to be treated separately, as individual games or game-like experiences, and as such it would be interesting to investigate how related crowdsourcing approaches might fit into the picture, as mentioned earlier. The question of combining the results of different crowdsourcing platforms—games or others—from a data and execution flow point of view also remains to be solved. The current state of the art in the area, at least as far as the Semantic Web community is concerned, is that more and more games with a purpose, most recently as Facebook applications, are being developed, but it is unclear how their results can be exploited beyond their original scope. This problem has been solved to some extent by openly publishing the generated data according to Linked Data principles. However, in order to further optimize the usability and usefulness of the data, one would need special-purpose metadata schemas that capture the most important parameters of the crowdsourcing project the data resulted from. This would enable the realization of so-called "crowdsourcing pipelines," which are essential in order to undertake almost any activity in the ontology life cycle using human computation, as these activities are too complex to be crowdsourced as they are. In addition, the question of choosing the most appropriate crowd-sourcing technique in terms of costs and quality of results becomes relevant, especially as GWAPs share many commonalities with approaches such as microtask platforms in terms of the tasks they can deal with. In [125] we conducted a small experiment comparing OntoPronto, our game for ontology development, and Amazon's Mechanical Turk, demonstrating the superiority of the latter. The experiments, however, also showed several principled and technical challenges associated with such comparisons. The former can mainly be traced back to the different set of incentives which operate in the two crowdsourcing instances and with the fact that the game we looked into was designed as a co-operation game, a model which cannot be easily mapped to the processes within Mechanical Turk. The second category of challenges is directly a consequence of the fact that tasks that are not fully decomposable into independent units—in our case, iterative tasks—are not well supported in Mechanical Turk. While the latter could be tackled through extensions of the conventional labor platform, some of which are already in development, the former could turn into an interesting area of research. The types of motivators that may make games successful can be very strong intrinsic drivers for high amounts contributions. Paired with consensual multi-player models which discourage cheating, they also offer a basic form of quality assurance. Realizing such a crowd-driven problem-solving environment with paid microtask platforms may lead to interesting effects in terms of the quality of the data collected, as microtask platforms face great challenges regarding spam prevention.

Knowledge Corpora

GAWPs usually need a corpus of knowledge to start with. This knowledge corpus is an integral part of the game challenges, because the challenges directly refer to entries in this corpus. Challenges cannot be shown too often to the same players without damaging the game experience [132]. This requires a large repository of knowledge in the background that can be used as input, whereas online collections of resources such as YouTube, Flickr, Wikipedia, or WordNet are surely also useful. In the context of SpotTheLink [123] we experimented with various ontologies in domains such as eCommerce and eTourism as knowledge corpus for the game. None of them received a positive resonance in initial trials. This was due to the structure and size of the ontologies (many inheritance levels, large number of concept siblings per level in the eCommerce setting) or the domain itself (perceived as less interesting by interviewees or simply too far away from their daily life). Ontologies suitable for an ontology alignment game should be of a manageable size (several hundreds of concepts) and in a domain that a broad audience of users can relate to (e.g., media, entertainment, sports, but also ontologies capturing general knowledge such as DBPedia and PROTON, as in SpotTheLink and WhoKnows?). These limitations may affect the acceptance of a game whose purpose is the learning of an ontology in a domain which is accessible to an expert audience. A second problem game designers have to solve is the quality of the knowledge corpus when it is generated automatically. In this case, some of the challenges presented to the users might not be meaningful, thus hampering the overall game experience.

Game Fun

The challenge here is to reach a balance between an appealing design and the purposefulness of the game with respect to the task to be solved. Games for semantic-content authoring are in many cases on the edge of being too difficult for a non-expert audience. The positive side of this is that such games provide an intellectual challenge, which is important to keep the games interesting.

The negative side is that creating an attractive and easy-to-grasp interface for such technical tasks is not easy; user interfaces studies for semantic technologies are still at the very beginning, even when it comes to expert-oriented environments such as ontology editors. In selection-agreement games, the navigation must force users to move along existing knowledge structures and make a consensual selection. This requires appropriate visualization techniques, while remaining "game-like" and "playful" in terms of the colors and metaphors used. An example of how this could be achieved is the Universe Game, however, at the expense of the usefulness and comprehensiveness of the collected annotations. Massive user participation and generation of output is crucial for the games and the methods they incorporate; as such, they require a critical mass of contributions. Even when the task is intellectually challenging, and the interface is perceived as usable and pleasant, we cannot expect a massive user involvement per se. Additional incentives schemes and motivation are needed to get users to play and to get them to continue playing. This can include competition, reputation, and sociability. Keeping scores is an important feature of every game. Players want to improve their ranking and standing within the community. Hence, score lists should be a big deal and progress should be rewarded through badges and similar things. Users should also be notified or made aware when they are about to lose a rank or when they have improved themselves and to provide means for users to be able to brag about the results on social platforms. With respect to sociability, knowing that they are playing against a real partner is also motivating for many players. As the games are cooperative by design, players might want to know more about their partner. Allowing communication after the gaming session—if they achieve a certain amount of points—could be such an incentive. Moreover, players should be able to indicate preferences for the choice of their partners, be able to invite people from their social network to play the game, and report extensively on their achievements, for instance, through frequent status updates.

Recent studies on the massive success of social games deployed via Facebook and other platforms, which are similar to games with a purpose through their relatively simple, casual plot and the incentive mechanisms they incorporate, have identified a number of essential game design features, which ensure a growth in the number of players and encourage player retention. Many of these features are available at best only in a very basic form in the games with a purpose we surveyed: multiple difficulty levels, changing game goals and virtual gifts, to name only a few. It is yet unclear how these features could be implemented in games aiming at solving a Semantic Web-related problem, as the underlying tasks are often repetitive, and distinguishing variable levels of difficulty and goals automatically is not straightforward. In most cases, the games have been evaluated in controlled experiments and there is scanted evidence on their acceptance by the large community of casual gamers as well as on their ability to achieve viral growth and players retention.

CHAPTER 6

Conclusions

Many tasks in the semantic data management life cycle are optimally approached as a combination of human and computational intelligence. The multitude of methodologies, techniques, and tools developed over the past decade addressing specific aspects of semantic data management offer many sophisticated features. Nevertheless, they are hardly accessible to a lay audience. This is due to the fact that they assume in-depth expertise not only with respect to the task at hand, but also with respect to the underlying processes and procedures according to which the tasks are executed. These limitations laid the foundations for the emergence of a new field of research in semantic technologies; this field was inspired by the success of the Web 2.0 phenomenon in encouraging user participation and capitalizing on the power of collective intelligence [17, 128, 134] (see also Chapter 1).

It uses games as an environment which attracts and retains users' attention and contributions, and tries to develop game scenarios in which during game play data are collected and used to improve automatic solutions to computationally hard problems. We have seen several examples of such games tackling tasks as diverse of ontology alignment, image and video annotation, and information extraction. They should illustrate the potential of the GWAP approach along a number of design challenges that should not be underestimated: the suitability of the task and knowledge domain, and the mechanisms set up to ensure high quality contributions and great user involvement through incentives and motivation.

Bibliography

[1] Participatory design of computer systems (panel). In Jeff Johnson, editor, *Proceedings of the SIGCHI conference on human factors in computing systems: Empowering people*, pages 141–144, 1990. 52

[2] F.E. Abadi, M.R. Jalilvand, M. Sharif, G.A. Salimi, and S.A. Khanzadeh. A study of influential factors on employees' motivation for participating in the in-service training courses based on modified expectancy theory. *International Business and Management*, 2(1):157–169, 2011. 19

[3] C. Abt. *Serious Games*. University Press of America, 2002. 68

[4] J.S. Adams. Inequity in social exchange. In L. Berkowitz, editor, *Advances in Experimental Social Psychology*, volume 2, pages 267–299. Academic Press, 1965. 24

[5] A.A Alchian and H. Demsetz. Production, Information Cost, and Economic Organization. *The American Economic Review*, (62):777–795, December 1972. 23

[6] A. G. Alfano and G. Marwell. Experiments on the provisions of public goods by groups iii. non-divisibility and free riding in "real" groups. *Social Psychology Quarterly*, 43(3):300–309, 1980. DOI: 10.2307/3033732 26

[7] Casual Games Association. Casual Games Market Report. 67 http://www.org.id.tue.nl/IFIP-TC14/documents/CasualGamesMarketReport-2007.pdf, 2007.

[8] Entertainment Software Association. Essential Facts About the Computer and Video Game Industry. http://www.theesa.com/facts/pdfs/ESA_Essential_Facts_2010.pdf, 2010. 67

[9] International Game Developers Association. 2008-2009 Casual Games White Paper. http://www.igda.org/casual/, 2009. 67

[10] A.G. Bedeian. *Management*. Dryden Press, New York, 3rd edition, 1993. 19

[11] S. Bloehdorn, K. Petridis, C. Saathoff, N. Simou, V. Tzouvaras, Y. Avrithis, S. Handschuh, Y. Kompatsiaris, S. Staab, and M. G. Strintzis. Semantic annotation of images and videos for multimedia analysis. In *Proceedings of the 2nd European Semantic Web Conference ESWC 2005*, pages 592–607, 2005. DOI: 10.1007/11431053_40 8

[12] Kerl Bodker, Finn Kensing, and Jesper Simonsen. *Participatory It Design: Designing for Business and Workplace Realities*. MIT Press, Cambridge, MA, USA, 2004. 51, 52

[13] S. Bodker, P. Ehn, J. Kammersgaard, M. Kyng, and Y. Sundblad. A UTOPIAN experience: On design of powerful computer-based tools for skilled graphic workers. *Computers and Democracy G. Bjerknes, P. Ehn, and M. Knyg (Eds.). Avebury Pub. England*, page 25, 1987. 52

[14] Barry W. Boehm. Software engineering economics. *Software Engineering, IEEE Transactions on*, SE-10(1):4 –21, January 1984. DOI: 10.1109/TSE.1984.5010193 52

[15] P. Bonacich, G. Shure, J. Kahan, and R. Meeker. Cooperation and Group Size in the N-Person Prisoner's Dilemma. *Journal of Conflict Resolution*, 20:687–706, 1976. DOI: 10.1177/002200277602000406 26

[16] E. Paslaru Bontas. *A Contextual Approach to Ontology Reuse: Methodology, Methods and Tools for the Semantic Web*. PhD thesis, Free University Berlin, 2007. 6

[17] S. Braun, A. Schmidt, A. Walter, G. Nagypal, and V. Zacharias. Ontology Maturing: A Collaborative Web 2.0 Approach to Ontology Engineering. In *Workshop on Social and Collaborative Construction of Structured Knowledge at the 16th International World Wide Web Conference WWW2007*, 2007. 13, 91

[18] I. Cantador, M. Fernández, and P. Castells. Improving Ontology Recommendation and Reuse in WebCORE by Collaborative Assessments. In *Proceedings of the 1st International Workshop on Social and Collaborative Construction of Structured Knowledge CKC 2007 at the at the 16th International World Wide Web Conference WWW 2007*, 2007. 9

[19] M.R. Carrel and J.E. Dittrich. Equity Theory: The Recent Literature, Methodological Considerations, and New Directions. *The Academy of Management Review*, 3(4):202–210, 1978. DOI: 10.2307/257661 24

[20] Jon Chamberlain, Massimo Poesio, and Udo Kruschwitz. A demonstration of human computation using the phrase detectives annotation game. In *Proceedings of the ACM SIGKDD Workshop on Human Computation*, pages 23–24, 2009. DOI: 10.1145/1600150.1600156 78

[21] Y. Chen, M. F. Harper, J. Konstan, and S. X. Li. Social Comparisons and Contributions to Online Communities: A Field Experiment on MovieLens. *American Economic Review*, 100(4):1358–1398, 2010. DOI: 10.1257/aer.100.4.1358 28

[22] R. Cuel, M. Herbrechter, M. Rohde, M. Stein, O. Tokarchuk, T. Wiedenhöer, F. Yetim, and M. Zamarian. Requirements Report of the INSEMTIVES Seekda! Use Case. Technical Report 1, International Institute for Socio-Informatics, 2011. 19

[23] R. Cuel, O. Morozova, M. Rohde, E. Simperl, K. Siorpaes, O. Tokarchuk, T. Widenhoefer, F. Yetim, and M. Zamarian. Motivation mechanisms for participation in human-driven semantic content creation. *Intl. J. of Knowledge Engineering and Data Mining*, 1(4):331–349, 2011. DOI: 10.1504/IJKEDM.2011.040653 19

[24] R. Cuel, O. Tokarchuk, and M. Zamarian. Mechanism Design for Designing Annotation Tools. In *Proceedings of the Sixth International Conference on Internet and Web Applications and Services IARIA2011*, 2011. 11

[25] Roberta Cuel, Marc Herbrechter, Markus Rohde, Martin Stein, Oksana Tokarchuk, Torben Wiedenhöfer, and Fahri Yetim. Requirements report of the INSEMTIVES seekda! use case. *International Reports on Socio-Informatics*, 8(1):4–20, 2011. 55

[26] R. M. Dawes, J. McTavish, and H. Shaklee. Behavior, Communication, and Assumptions About Other People's Behavior in a Commons Dilemma Situation. *Journal of Personality and Social Psychology*, 35(1–11), 1977. DOI: 10.1037/0022-3514.35.1.1 27

[27] R. M. Dawes, H. Shaklee, and F. Talarowski. On getting people to cooperate when facing a social dilemma: Moralizing helps. Technical Report 76–82, Oregon Research Institute Eugene, 1976. 28

[28] R.M. Dawes. Social Dilemmas, Economic Self-Interest, and Evolutionary Theory. In D.R. Brown and J.E.K. Smith, editors, *Frontiers of Mathematical Psychology: Essays in Honor of Clyde Coombs*, pages 53–79. Springer Verlag, 1991. 27

[29] E. L. Deci and R. M. Ryan. The "What" and "Why" of Goal Pursuits: Human Needs and the Self-Determination of Behavior. *Psychological Inquiry*, 11:227–268, 2000. DOI: 10.1207/S15327965PLI1104_01 19, 22

[30] M. Dimitrov, A. Simov, V. Momtchev, and M. Konstantinov. WSMO Studio - a Semantic Web Services Modelling Environment for WSMO (System Description). In *Proceedings of the 4th European Semantic Web Conference ESWC2007*, 2007. DOI: 10.1007/978-3-540-72667-8_53 8

[31] R. Drago and G.T. Garvey. Incentives for Helping on the Job: Theory and Evidence. *Journal of Labor Economics*, 16(1):1–25, 1998. DOI: 10.1086/209880 23

[32] B. Norton E. Simperl and D. Vrandecic. Crowdsourcing tasks in Linked Data management. In *Proceedings of the 2nd workshop on consuming Linked Data COLD 2011 co-located with the 10th International Semantic Web Conference ISWC 2011*, 2011. 61

[33] Pelle Ehn. Scandinavian design: on participation and skill. In Paul S. Adler and Terry A. Winograd, editors, *Usability*, pages 96–132. Oxford University Press, Inc., New York, NY, USA, 1992. 52

[34] J. Euzenat, A. Mocan, and F. Scharffe. *Ontology Alignment*. Springer, 2007. 9

[35] J. Euzenat and P. Shvaiko. *Ontology Matching*. Springer, 2007. 9

[36] S. M. Falconer and M.-A. Storey. A cognitive support framework for ontology mapping. In *Proceedings of the 6th International Semantic Web Conference (ISWC)*, pages 114–127, 2007. DOI: 10.1007/978-3-540-76298-0_9 9

[37] J. Fox and M. Guyer. Group Size and Others' Strategy in an N-Person Game. *Journal of Conflict Resolution*, 21:323–338, 1977. DOI: 10.1177/002200277702100206 26

[38] J. Fox and M. Guyer. Public Choice and Cooperation in N-Person Prisoner's Dilemma. *Journal of Conflict Resolution*, 22:468–481, 1978. DOI: 10.1177/002200277802200307 28

[39] B.S. Frey and S. Meier. Social Comparisons and Pro-social Behavior Testing Conditional Cooperation in a Field Experiment. *American Economic Review*, 94:1717–1722, 2004. DOI: 10.1257/0002828043052187 28

[40] F. Fukuyama. *Trust: The Social Virtues and the Creation of Prosperity*. The Free Press, 1995. 22

[41] N.S. Glance and B.A. Huberman. The Dynamics of Social Dilemmas. *Scientific American*, pages 76–81, March 1994. DOI: 10.1038/scientificamerican0394-76 27

[42] L. Gomes. Will All of Us Get Our 15 Minutes On a YouTube Video? Wallstreet Journal, 2006. 13

[43] A. Gómez-Pérez. Evaluation of Ontologies. *International Journal of Intelligent Systems*, 16(3):391–409, 2001. DOI: 10.1002/1098-111X(200103)16:3%3C391::AID-INT1014%3E3.0.CO;2-2 9

[44] A. Gomez-Perez, M. Fernandez-Lopez, and O. Corcho. *Ontological Engineering*. Advanced Information and Knowledge Processing. Springer, 2004. 3, 4, 5

[45] Carl Goodman. Virtual World Content Annotation Prototype. Deliverable D7.2.2 of the project FP7-ICT-2007-3-231181 INSEMTIVES, 2011. 82

[46] M. Granovetter. Economic action and social structure: The problem of embeddedness. *American Journal of Sociology*, 91:481–510, 1985. DOI: 10.1086/228311 22

[47] Joan M. Greenbaum and M. Kyng, editors. *Design at Work: Cooperative Design of Computer Systems*. L. Erlbaum Associates Inc., Hillsdale, NJ, USA, 1991. 52

[48] R.W. Griffin. *Task Design: An Integrative Approach*. Scott, Foresman and Company, Glenview, IL, 1982. 22

[49] T. R. Gruber. Toward Principles for the Design of Ontologies Used for Knowledge Sharing. *International Journal of Human-Computer Studies*, 43:907–928, 1995. DOI: 10.1006/ijhc.1995.1081 3

[50] N. Guarino and C. A. Welty. Evaluating ontological decisions with ontoclean. *Communications of the ACM*, 45(2):61–65, 2002. DOI: 10.1145/503124.503150 9

[51] J. Hackman and G.R. Oldham. *Work redisign*. Addison-Wesley, 1980. 19, 22

[52] C. Hansson, Dittrich Y., and D. Randall. The development is driven by our users, not by ourselves" - including users in the development of an off-the-shelf software. In *26th Information Systems Research Seminar in Scandinavia (IRIS 26)*, 2003. 52

[53] Marc Hassenzahl. The thing and i: understanding the relationship between user and product. In *Funology: from usability to enjoyment*, pages 31–42. Kluwer Academic Publishers, 2004. 59

[54] T. Heath and C. Bizer. *Linked Data: Evolving the Web into a Global Data Space*. Synthesis Lectures on the Semantic Web Theory and Technology. Morgan & Claypool, 2011. 3, 8

[55] A. Hemetsberger. When consumers produce on the internet: The relationship between cognitive-affective, socially-based, and behavioral involvement of prosumers. *The Journal of Social Psychology*, 2003. 12

[56] Marc Herbrechter. Online participatory design - implementierung einer webanwendung zur unterstuetzung verteilt-partizipativer softwareentwicklung. 2011. 56

[57] Marleen Huysman, Etienne Wenger, and Volker Wulf. *Communities and technologies: proceedings of the first International Conference on Communities and Technologies, C&T 2003*. Springer, Dordrecht, September 2003. 51

[58] Marleen Huysman and Volker Wulf. *Social capital and information technology*. June 2004. 51

[59] P. Ipeirotis, F. Provost, and J. Wang. Quality management on Amazon Mechanical Turk. In *Proceedings of the ACM SIGKDD Workshop on Human Computation*, pages 64–67, 2010. DOI: 10.1145/1837885.1837906 61, 73

[60] R.M. Isaac, D. Schmidtz, and J.M. Walker. The Assurance Problem in a Laboratory Market. *Public Choice*, 63(3):217–236, 1988. DOI: 10.1007/BF02337743 27

[61] R.M. Isaac, J. Walker, and A. Williams. Group Size and the Voluntary Provision of Public Goods: Experimental Evidence Utilizing Very Large Groups. 1990. DOI: 10.1016/0047-2727(94)90068-X 26, 27

[62] M. Jensen and W.H. Meckling. Theory of the Firm: Managerial Behavior, Agency Costs and Ownership Structure. *Journal of Financial Economics*, 3(4):305–360, 1976. DOI: 10.1016/0304-405X(76)90026-X 23

[63] D.C. Jones and T. Kato. The Productivity Effects of Stock Ownership Plans and Bonuses: Evidence from Japanese Panel Data. *American Economic Review*, 85:391–414, 1995. 23

[64] Steven J. Karau and Kipling D. Williams. Social loafing: A meta-analytic review and theoretical integration. *Journal of Personality and Social Psychology*, 65(4):681–706, 1993. DOI: 10.1037/0022-3514.65.4.681 52

[65] Gabriella Kazai, Natasa Milic-Frayling, and Jamie Costello. Towards methods for the collective gathering and quality control of relevance assessments. In *Proceedings of the 32nd international ACM SIGIR conference on research and development in information retrieval*, pages 452–459, 2009. DOI: 10.1145/1571941.1572019 69

[66] H.H. Kelley and J. Grzelak. Conflict Between Individual and Common Interest in an N-Person Relationship. *Journal of Experimental Social Psychology*, 21:190–197, 1972. DOI: 10.1037/h0032224 26

[67] Finn Kensing and Jeanette Blomberg. Participatory Design: Issues and Concerns. *Computer Supported Cooperative Work (CSCW)*, 7(3):167–185, 1998. DOI: 10.1023/A:1008689307411 52

[68] N. Kerr. Illusions of Efficacy: The Effects of Group Size on Perceived Efficacy in Social Dilemmas. *Journal of Experimental Social Psychology*, 25:287–313, 1989. DOI: 10.1016/0022-1031(89)90024-3 27

[69] M. Kerrigan, A. Mocan, E. Simperl, and D. Fensel. Modeling semantic web services with the web service modeling toolkit. Technical report, Semantic Technology Institute (STI), 2008. 8

[70] M. Kerrigan, A. Mocan, M. Tanler, and D. Fensel. The web service modeling toolkit - an integrated development environment for semantic web services (system description). In *European Semantic Web Conference (ESWC 2007), Innsbruck, Austria*, 2007. DOI: 10.1007/978-3-540-72667-8_57 8

[71] A. Kittur, E. Chi, and B. Suh. Crowdsourcing user studies with Mechanical Turk. In *Proceddings of the 26th annual SIGCHI conference on human factors in computing systems*, pages 453–456, 2008. DOI: 10.1145/1357054.1357127 61

[72] A. Kittur, E.H. Chi, and B. Suh. 7crowdsourcing user studies with mechanical turk. DOI: 10.1145/1357054.1357127 61

[73] M. Knez and D. Simester. Firm-wide Incentives and Mutual Monitoring at Continental Airlines. *Journal of Labor Economics*, 19(4):743–772, 2001. DOI: 10.1086/322820 23

[74] S. Komorita and S. Lapworth. Cooperative Choice among Individuals Versus Groups in an N-Person Dilemma. *Journal of Personality and Social Psychology*, 42:487–496, 1982. DOI: 10.1037/0022-3514.42.3.487 26

[75] S. Kuznetsov. Motivations of contributors to wikipedia. *ACM SIGCAS Computers and Society*, 36(2), 2006. DOI: 10.1145/1215942.1215943 11, 12, 13, 24

[76] K.R. Lakhani and R.G. Wolf. Why hackers do what they do: Understanding motivation and effort in free/open source software projects. *In Perspectives on Free and Open Source Software, MIT Press*, 2005. 19

[77] Edith Law and Luis von Ahn. Input-agreement: a new mechanism for collecting data using human computation games. In *Proceedings of the 27th international conference on human factors in computing systems CHI 2009*, pages 1197–1206, 2009. DOI: 10.1145/1518701.1518881 14, 72, 73

[78] J. Ledyard. Public goods: A survey of experimental research. In J. Kagel and A. Roth, editors, *Handbook of Experimental Economics*, pages 111–194. Princeton University Press, 1995. 26, 27

[79] W.B.G. Liebrand. The effect of social motives, communication and group size on behaviour in an n-person multi-stage mixed-motive game. *European Journal of Social Psychology*, 14:239–264, 1984. DOI: 10.1002/ejsp.2420140302 26, 27

[80] Siegwart Lindenberg. Intrinsic motivation in a new light. *Kyklos*, 54(2-3):317–324, 2001. DOI: 10.1111/1467-6435.00156 19

[81] E.A. Locke and G.P. Latham. *A Theory of Goal Setting and Task Performance*. Prentice Hall, 1990. 21

[82] J. Lopez and J. Scott. *Social Structure*. Open University Press, 2000. 22

[83] A. Lozano-Tello and A. Gomez-Perez. Ontometric: A method to choose the appropriate ontology. *Journal of Database Management*, 15(2), 2004. DOI: 10.4018/jdm.2004040101 9

[84] m. c. schraefel and Lloyd Rutledge, editors. *Special Issue User Interaction in Semantic Web Research*, volume 8(4) of *Journal of Web Semantics*, 2010. 5

[85] A. Maedche and S. Staab. Ontology Learning for the Semantic Web. *IEEE Intelligent Systems*, 16(2):72–79, 2001. DOI: 10.1109/5254.920602 9

[86] T. W. Malone. What makes things fun to learn? Heuristics for designing instructional computer games. In *Proceedings of the 3rd ACM SIGSMALL Symposium and the first SIGPC Symposium on Small Systems*, SIGSMALL '80, pages 162–169, 1980. DOI: 10.1145/800088.802839 67

[87] Michael Mandel and Daniel Ellis. A Web-Based Game for Collecting Music Metadata. *Journal of New Music Research*, 37(2):151–165, 2008. DOI: 10.1080/09298210802479300 69

[88] Cameron Marlow, Mor Naaman, Danah Boyd, and Marc Davis. Position Paper, Tagging, Taxonomy, Flickr, Article, ToRead. In *Proceedings of the Collaborative Web Tagging Workshop at WWW 2006*, 2006. 12

[89] G. Marwel and P. Oliver. *The Critical Mass in Collective Action: A Micro-Social Theory*. Cambridge University Press, 1993. DOI: 10.1017/CBO9780511663765 27

[90] P. Mika. Ontologies are us: A unified model of social networks and semantics. *Journal of Web Semantics*, 5(1):5–15, 2007. DOI: 10.1016/j.websem.2006.11.002 13

[91] monica schraefel, Daniel Alexander Smith, Igor Popov, Max Van Kleek, and Nigel Shadbolt. Will this work for Susan? Challenges for Delivering Usable and Useful Generic Linked Data Browsers. Technical report, University of Southampton, June 2010. 5

[92] Anders I. Mørch and Nikolay D. Mehandjiev. Tailoring as collaboration: The mediating role of multiple representations and application units. *Computer Supported Cooperative Work (CSCW)*, 9(1):75–100, 2000. DOI: 10.1023/A:1008713826637 52

[93] N. Fridman Noy and M. Musen. The prompt suite: Interactive tools for ontology merging and mapping. *International Journal of Human Computer Studies*, 59(6):983–1024, 2003. DOI: 10.1016/j.ijhcs.2003.08.002 9

[94] A. Ngonga Ngomo and Sören Auer. LIMES A Time-Efficient Approach for Large-Scale Link Discovery on the Web of Data. In *Proceedings of the Proceedings of the 22nd International Joint Conference on Artificial Intelligence IJCAI2011*, pages 2312–2317, 2011. DOI: 10.5591/978-1-57735-516-8/IJCAI11-385 8

[95] Axel. Ngonga Ngomo and K. Lyko. EAGLE: Efficient Active Learning of Link Specifications Using Genetic Programming. In *Proceedings of the 9th European Semantic Web Conference ESWC2012*, pages 149–163, 2012. DOI: 10.1007/978-3-642-30284-8_17 8

[96] N. Noy and M. Musen. Anchor-PROMPT: Using Non-Logical Context for Semantic Matching. In *Proceedings of the IJCAI Workshop on Ontologies and Information Sharing*, pages 63–70, 2001. 9

[97] J. Orbell, R. Dawes, and A. Van de Kragt. The Limits of Multilateral Promising. *Ethics*, 100:616–627, 1990. DOI: 10.1086/293213 27

[98] Wanda J. Orlikowski. The duality of technology: Rethinking the concept of technology in organizations. *Organization Science*, 3(3):398–427, January 1992. DOI: 10.1287/orsc.3.3.398 52

[99] P. Cimiano P. Buitelaar and B. Magnini, editors. *Ontology Learning from Text: Methods, Evaluation And Applications*. IOP Press, 2005. 7

[100] H.S. Pinto and J.P. Martins. A methodology for ontology integration. In *International Conference on Knowledge Capture (K-CAP)*, pages 131–138. ACM Press, 2001. DOI: 10.1145/500737.500759 9

[101] Igor Popov. mashpoint: Supporting Data-centric Navigation on the Web. In *Proceedings of the 2012 ACM Annual Conference Extended Abstracts on Human Factors in Computing Systems CHI2012*, pages 2249–2254, 2012. DOI: 10.1145/2212776.2223784 5

[102] Jenny Preece. *Online Communities: Designing Usability, Supporting Sociability: Supporting Sociability, Designing Usability*. John Wiley & Sons, 1. auflage edition, August 2000. 51

[103] C. Prendergast. The Provision of Incentives in Firms. *Journal of Economic Literature*, 37(1):7–63, 1999. DOI: 10.1257/jel.37.1.7 23

[104] A. Rapoport, G. Bornstein, and I. Erev. Intergroup Competition for Public Goods: Effects of Unequal Resource and Relative Group Size. *Journal of Personality and Social Psychology*, 56(5):748–756, 1989. DOI: 10.1037/0022-3514.56.5.748 27

[105] L. Reeve and H. Han. *Survey of Semantic Annotation Platforms*, pages 1634–1638. ACM Press, 2005. 7

[106] S.A. Ross. The Economic Theory of Agency: The Principal's Problem. *American Economic Review*, LXII:134–139, 1973. 23

[107] W. W. Royce. Managing the development of large software systems : Concepts and techniques. In *Technical Papers of Western Electronic Show and Convention*, Los Alamitos, CA, USA, 1970. IEEE Computer Society. 52

[108] m.c. schraefel, Jennifer Golbeck, Duane Degler, Abraham Bernstein, and Lloyd Rutledge. Semantic Web User Interactions: Exploring HCI Challenges. In *Proceedings of the 2008 ACM Annual Conference Extended Abstracts on Human Factors in Computing Systems CHI2008*, pages 3929–3932, 2008. DOI: 10.1145/1358628.1358959 5

[109] Douglas Schuler and Aki Namioka. *Participatory Design: Principles and Practices*. Routledge, March 1993. 52

[110] L. Seneviratne and E. Izquierdo. An interactive framework for image annotation through gaming. In *Proceedings of the International Conference on Multimedia Information Retrieval MIR 2010*, pages 517–526, 2010. DOI: 10.1145/1743384.1743473 69, 72

[111] H.A. Simon. Motivational and Emotional Controls of Cognition. In *Models of Thought*, pages 29–38. Yale University Press, 1967. 23

[112] E. Simperl, S. Wölger, B. Norton, S. Thaler, and T. Bürger. Combining human and computational intelligence: the case of data interlinking tools. *International Journal of Metadata, Semantics and Ontologies*, 2012. DOI: 10.1504/IJMSO.2012.050018 9

[113] K. Siorpaes. *Games for Semantic Content Creation*. PhD thesis, STI Innsbruck, University of Innsbruck, 2009. 14

[114] K. Siorpaes and M. Hepp. Games with a purpose for the semantic web. *IEEE Intelligent Systems*, 23(3):50–60, 2008. DOI: 10.1109/MIS.2008.45 11, 69

[115] K. Siorpaes and E. Simperl. Human intelligence in the process of semantic content creation. *World Wide Web Journal*, 13(1):33–59, 2010. DOI: 10.1007/s11280-009-0078-0 5, 8, 9

[116] K. Siorpaes and S. Thaler. Requirements and design of a generic gaming toolkit and API. Deliverable D4.1.1 of the project FP7-ICT-2007-3-231181 INSEMTIVES, 2010. 84

[117] I. D. Steiner. *Group Process and Productivity (Social Psychological Monograph)*. Academic Press Inc, 1972. 86

[118] H. Tajfel. *Human Groups and Social Categories*. Cambridge University Press, 1981. 27

[119] H. Tajfel and J.C. Turner. An Integrative Theory of Intergroup Conflict. In W.G. Austin and S. Worchel, editors, *The Social Psychology of Intergroup Relations*. Brooks-Cole, 1979. 24

[120] C. Tempich, E. Simperl, M. Luczak, and H. Pinto. Argumentation-Based Ontology Engineering. *IEEE Intelligent Systems*, 22(6):52–59, 2007. DOI: 10.1109/MIS.2007.103 13

[121] I. Terziev, A. Kiryakov, and D. Manov. Base-upper-level ontology guidance. Deliverable D1.8.1. of the EU-IST project IST-2003-506826 SEKT, 2005. 74

[122] S. Thaler. Generic Gaming Toolkit and API implementation. Deliverable D4.1.2 of the project FP7-ICT-2007-3-231181 INSEMTIVES, 2010. 84

[123] S. Thaler, E. Simperl, K. Siorpaes, and S. Wölger. *Collaboration and the Semantic Web: Social Networks, Knowledge Networks and Knowledge Resources*, chapter SpotTheLink: A Game-based Approach to the Alignment of Ontologies, pages 40–63. IGI Global, 2012. 69, 74, 88

[124] S. Thaler, K. Siorpaes, C. Hofer, and E. Simperl. A survey on games for knowledge acquistion. Technical report, STI Innsbruck, University of Innsbruck, May 2011. 74, 80

[125] Stefan Thaler, Elena Simperl, and Stephan Wolger. An experiment in comparing human-computation techniques. *IEEE Internet Computing*, 99(PrePrints), 2012. DOI: 10.1109/MIC.2012.67 68, 87

[126] J. Thom-Santelli, M.J. Muller, and D.R. Millen. Social Tagging Roles: Publishers, Evangelists, Leaders. In *Proceedings of the 26th Annual ACM Conference on Human Factors in Computing Systems*, pages 1041–1044, 2008. DOI: 10.1145/1357054.1357215 19

[127] M. Toda. Emotions Viewed as Tightly Organized, Genetically Determined System of Behaviour-Selection Programs. In J.T. Spence and C.E. Izard, editors, *Motivation, Emotion, and Personality*, pages 261–273. Elsevier Science Publishers, 1985. 24

[128] Tania Tudorache, Sean M. Falconer, Natalya Fridman Noy, Csongor Nyulas, Tevfik Bedirhan Üstün, Margaret-Anne D. Storey, and Mark A. Musen. Ontology development for the masses: Creating icd-11 in webprotégé. In *Knowledge Engineering and Management by the Masses - Proceedings of the 17th International Conference (EKAW 2010)*, pages 74–89, 2010. 91

[129] V. Uren, P. Cimiano, J. Iria, S. Handschuh, M. Vargas-Vera, E. Motta, and F. Ciravegna. Semantic annotation for knowledge management: Requirements and a survey of the state of the art. *Web Semantics: Science, Services and Agents on the World Wide Web*, 4(1):14–28, 2006. DOI: 10.1016/j.websem.2005.10.002 7

[130] Victoria Uren, Yuangui Lei, Vanessa Lopez, Haiming Liu, Enrico Motta, and Marina Giordanino. The usability of semantic search tools: A review. *Knowledge Engineering Review*, 22(4):361–377, 2007. DOI: 10.1017/S0269888907001233 5

[131] M. Uschold and M. King. Towards a methodology for building ontologies. In *Workshop on Basic Ontological Issues in Knowledge Sharing*, Montreal, Canada, 1995. 6

[132] L. Van Ahn and L. Dabbish. Designing games with a purpose. *Communications of the ACM*, 51(8):58–67, 2008. DOI: 10.1145/1378704.1378719 11, 14, 69, 73, 74, 80, 88

[133] David Vickrey, Aaron Bronzan, William Choi, Aman Kumar, Jason Turner-Maier, Arthur Wang, and Daphne Koller. Online word games for semantic data collection. In *Proceedings of the Conference on Empirical Methods in Natural Language Processing*, pages 533–542, 2008. DOI: 10.3115/1613715.1613781 69, 72

[134] Max Völkel, Markus Krötzsch, Denny Vrandecic, Heiko Haller, and Rudi Studer. Semantic Wikipedia. In *Proceedings of the 15th international conference on World Wide Web WWW 2006*, pages 585–594, 2006. DOI: 10.1145/1135777.1135863 13, 91

[135] Johanna Völker, Denny Vrandecic, York Sure, and Andreas Hotho. Learning disjointness. In *Proceedings of the 4th European Semantic Web Conference ESWC 2012*, pages 175–189, 2007. DOI: 10.1007/978-3-540-72667-8_14 7

[136] Julius Volz, Christian Bizer, Martin Gaedke, and Georgi Kobilarov. Discovering and maintaining links on the web of data. In *Proceedings of the 8th International Semantic Web Conference ISWC 2009*, pages 650–665, 2009. DOI: 10.1007/978-3-642-04930-9_41 8

[137] L. von Ahn. Games with a Purpose. *Computer*, 39(6):92–94, 2006. 67

[138] Luis von Ahn, Ruoran Liu, and Manuel Blum. Peekaboom: a game for locating objects in images. In *Proceedings of the SIGCHI conference on human factors in computing systems CHI 2006*, pages 55–64, 2006. DOI: 10.1145/1124772.1124782 73

[139] E. Von Hippel. Horizontal innovation networks - by and for users. Working Paper 4366-02, 2002. DOI: 10.1093/icc/dtm005 19

[140] C. Wagner and P. Prasarnphanich. Innovating Collaborative Content Creation: The Role of Altruism and Wiki Technology. In *Proceedings of the 40th Annual Hawaii International Conference on System Science*, pages 18–27, 2007. DOI: 10.1109/HICSS.2007.277 19

[141] S. Wasko, M. Faraj. Why Should I Share? Examining Social Capital and Knowledge Contribution in Electronic Networks of Practice. *MIS Quarterly*, 29(1):25–57, 2005. 19

[142] D. White. Results and analysis of web 2.0 services survey, 2007. 19

[143] C. Wiertz and K. De Ruyter. Beyond the Call of Duty: Why Customers Contribute to Firm-Hosted Commercial Online Communities. *Organization Studies*, 28(3):347–376, 2007. DOI: 10.1177/0170840607076003 19

[144] S. Wölger, K. Siorpaes, T. Bürger, E. Simperl, S. Thaler, and C. Hofer. Interlinking data - approaches and tools. Technical report, STI Innsbruck, University of Innsbruck, March 2011. 9

[145] T. Yamagishi and K.S. Cook. Generalized exchange and social dilemmas. *Social Psychology Quarterly*, 56:235–248, 1993. DOI: 10.2307/2786661 26

[146] E. Zimmerman and K. Salen. *Rules of Play*. MIT Press, 2004. 68

Authors' Biographies

ELENA SIMPERL

Elena Simperl works as a Senior Lecturer at the University of Southampton in the UK. She has been active as a Semantic Web researcher for almost a decade, authoring over 75 publications, and serving as Chair in several relevant scientific conferences, such as the European Semantic Web Conference, which she coordinated as General Chair in 2012. Elena has been involved in over 20 European and national projects in the field of semantic technologies. Among others she led the European research project INSEMTIVES, which dealt with questions related to this book.

ROBERTA CUEL

Roberta Cuel is an Assistant Professor of Organization Studies in the Faculty of Economics, University of Trento in Italy. Her research interests aim at discovering the interdependencies between technology and organizations, in particular the impacts of innovative technologies on teams, communities, and organizational models, the study of distributed tools and processes which allow organizational learning and knowledge management, and knowledge representation systems as mechanisms for knowledge reification processes. She has a PhD in organization and management from the University of Udine and has written a number of chapters in books, as well as articles in international journals and has served as a Program Committee member for various interdisciplinary conferences.

MARTIN STEIN

Martin Stein studied Information Systems at the University of Siegen in Germany. During his studies he worked on different nationally and EU-funded projects. He wrote his diploma thesis on the topic of "Motivation for Participation - Meta Data Creation for the Eclipse Ecosystem". Since 2010 he has been working as a research associate at the Institute for Information Systems and New Media at the University of Siegen. His research interests are HCI, ambient assisted living and design for the aging society.